#17

D0403067

Chaos in
the Cosmos

The Stunning Complexity
of the Universe

OTHER RECOMMENDED BOOKS
BY BARRY PARKER

STAIRWAY TO THE STARS
The Story of the World's Largest Observatory

THE VINDICATION OF THE BIG BANG
Breakthroughs and Barriers

COSMIC TIME TRAVEL
A Scientific Odyssey

COLLIDING GALAXIES
The Universe in Turmoil

INVISIBLE MATTER AND THE
FATE OF THE UNIVERSE

CREATION
The Story of the Origin and Evolution of the Universe

SEARCH FOR A SUPERTHEORY
From Atoms to Superstrings

EINSTEIN'S DREAM
The Search for a Unified Theory of the Universe

Chaos in the Cosmos

The Stunning Complexity of the Universe

BARRY PARKER

PLENUM PRESS • NEW YORK AND LONDON

Library of Congress Cataloging-in-Publication Data

Parker, Barry R.
    Chaos in the cosmos : the stunning complexity of the universe /
  Barry Parker.
       p.   cm.
    Includes bibliographical references and index.
    ISBN 0-306-45261-8 (alk. paper)
    1. Astrophysics.   2. Chaotic behavior in systems.   I. Title.
  QB43.2.P33   1996
  520'.151474--dc20   96-1482
                                                                  CIP

ISBN 0-306-45261-8

© 1996 Barry Parker
Plenum Press is a Division of Plenum Publishing Corporation
233 Spring Street, New York, N.Y. 10013-1578

10 9 8 7 6 5 4 3 2 1

Printed in the United States of America

Preface

The year was 1889. The French physicist-mathematician Henry Poincaré could not believe his eyes. He had worked for months on one of the most famous problems in science—the problem of three bodies moving around one another under mutual gravitational attraction—and what he was seeing dismayed and troubled him. Since Newton's time it had been assumed that the problem was solvable. All that was needed was a little ingenuity and considerable perseverance, but Poincaré saw that this was not the case. Strange, unexplainable things happened when he delved into the problem; it was not solvable after all. Poincaré was shocked and dismayed by the result—so disheartened he left the problem and went on to other things.

What Poincaré was seeing was the first glimpse of a phenomenon we now call chaos. With his discovery the area lay dormant for almost 90 years. Not a single book was written about the phenomenon, and only a trickle of papers appeared. Then, about 1980 a resurgence of interest began, and thousands of papers appeared along with dozens of books. The new science of chaos was born and has attracted as much attention in recent years as breakthroughs in superconductivity and superstring theory.

What is chaos? Everyone has an impression of what the word means, but scientifically chaos is more than random behavior, lack of control, or complete disorder. In this book I will attempt to explain what chaos is and what the excitement is all about. Chaos theory, as you will see, has a rich and fascinating history; it is an exciting new science, one that may take its place beside the great theories of our time. It is, however, a controversial theory. It is young, and we are still uncertain of many things, but it shows considerable promise. Many new insights into the universe may come from it.

The thrust of this book is chaos in astronomy, so chaos in black holes, pulsating stars, colliding galaxies, and the formation of the universe will be a central feature, but fractals, strange attractors, stretching and folding space, Julia sets, Mandelbrot sets, and other topics of chaos will also be discussed.

It is difficult in a book such as this to avoid technical terms completely. I have avoided them as much as possible, and have tried to explain them when they appear. For anyone who needs it, a glossary is supplied at the back of the book.

I am grateful to the scientists who assisted me. Interviews were conducted, mostly by telephone, with many of the people mentioned in the book. In some cases they supplied me with photographs and reprints. I would like to express my gratitude to them. They are Beverly Berger, Luca Bombelli, Robert Buchler, Matthew Choptuik, Matthew Collier, George Contoupolos, Martin Duncan, Jerry Golub, David Hobill, George Irwin, K. A. Innanen, James Lochner, Terry Matilsky, Vincent Moncrief, Svend Rugh, Gerald Sussman, Jean Swank, J. Wainright, and Jack Wisdom.

Most of the drawings were done by Lori Scoffield. I would like to thank her for an excellent job. I would like to thank Matthew Collier and George Irwin for several computer-generated diagrams. I would also like to thank Linda Greenspan Regan, Melicca McCormick, and the staff of Plenum for their help in bringing the book to its final form. Finally I would like to thank my wife for her support while the book was being written.

Contents

1

Introduction

A *solar eclipse is an awe-inspiring event. The disk of the moon cuts off more and more of the sun. The sky begins to darken, then* total darkness descends as the corona bursts into view. What is equally amazing about an eclipse is that we can predict the exact time at which it will occur years in advance. In fact we can predict to an exceedingly high accuracy the positions of all the planets for years into the future. It's easy to see, however, that this is not the case with all phenomena in nature. All you have to do is look upward at the clouds drifting overhead. As you watch them break up and reform, try to predict what will happen to a small section; you will soon find that most of the time you are wrong. The changes that take place are random. If you watch a leaf fall from a tree, you see it sway back and forth in the wind as it falls. If you try calculate its zigzagging path to the ground, you will soon find you can't; it's an impossible task.

It seems strange that we can predict things like the orbits of the planets so accurately, while others, like the falling leaf, seem to be beyond us. Yet both obey the same laws of nature. Gravity guides the planets in their course, and also attracts the leaf to Earth.

The motions of the planets and most other dynamic phenomena on Earth are described by Newton's laws, or more generally what is called classical mechanics. If we know the initial position of an object in a force field such as gravity we can, in theory, predict how it will behave in the future. This is what is referred to as determinism: The past determines the future.

At one time scientists were convinced that everything in the universe could be predicted if enough computing power was available. In other words, all of nature is deterministic. We now know, however, that this isn't true. Something as simple as the trajectory of a twig floating in a cascading mountain stream is beyond us. And, as you might expect, long-term prediction of the weather is also impossible.

Even things that we might think of as deterministic can become chaotic under certain circumstances. The simple pendulum—with a motion so predictable it has been used as the basis of clocks for centuries—can become chaotic. Make the bob of iron and place two magnets below it, and its motion will quickly become erratic.

Most of us use the word "chaos" rather loosely to represent anything that occurs randomly, so it is natural to think that the motion described by the erratic pendulum above is completely random. Not so. The scientific definition of chaos is different from the one you may be used to in that it has an element of determinism in it. This might seem strange, as determinism and chaos are opposites of one another, but oddly enough they are also compatible.

Scientifically, chaos is defined as extreme sensitivity to initial conditions. If a system is chaotic, when you change the initial state of the system by a tiny amount you change its future significantly. For example, if you start a twig in a stream at one point, then start it at another point only a few inches away, the two paths will be completely different. There will, in fact, be no resemblance between them. How is this deterministic? It is deterministic in that the chaos is governed by a law, and as we will see this law gives it structure and order.

Although chaos and chaotic phenomena have been known for centuries, determinism dominated science until the middle of the 19th century. It was obvious, however, that some systems were deterministic in theory only; they were far too complex to be truly deterministic. A good example is a gas; it is composed of molecules, each undergoing thousands of collisions every second. In theory you could calculate all the properties of a gas if you could follow each particle, that is, determine each of their trajectories. But it goes without saying that this would be impossible. Scientists eventually got around this with the invention of a new science: statistical mechanics. Based on probability theory, statistical mechanics only gives average values, but they were all that is needed. It can completely describe the bulk behavior of an assembly of particles such as a gas.

With the development of statistical mechanics there were two theories; one for determining the motions of simple systems consisting of few objects, and one for complex systems. Both worked well, but there was no apparent relationship between them; mathematically, they were completely different. Then in the late 1920s a third theory emerged. Like statistical mechanics it was based on probabilities; the orbit of an electron, for example, could be determined using a "probability wave." Called quantum mechanics, it applied only to the atomic world.

Scientists now had three theories at their disposal, and strangely, none of them could give us the trajectory of a twig on a mountain stream, or the path of the smoke rising from a campfire. But the seeds of a theory that would eventually help us understand these phenomena had been planted. It was not until the 1960s and 1970s, however, that it really started to grow. This new science was chaos theory. As it grew and matured, scientists saw that it was universal. It applied not only to physical problems such as the trajectories of objects in turbulent media (e.g, the twig in the swirling stream), but also to problems in biology, mathematics, chemistry, engineering, medicine, astronomy, and even business. Many of the most important discoveries of chaos theory, in fact, were made independently in different disciplines.

Chaos theory is now considered by many to be one of the most important discoveries of the 20th century, ranking alongside quantum theory and the theory of relativity. What is particularly exciting about the theory is that it has shown that chaos is not restricted to complex systems. In fact, many simple systems, even systems described by a single variable, can behave chaotically.

The theory began as a number of ideas that seemed to have little or no connection. But gradually scientists and mathematicians began to realize that each idea was part of a larger scheme—a theory that could explain phenomena in many different areas.

Some of the most important techniques of chaos theory can be traced back to the French physicist and mathematician, Henri Poincaré who lived and worked near the turn of the century. Poincaré is generally acknowledged to be the last universalist in mathematics, a man who was not only expert in all branches of mathematics, but also capable of making important contributions to any of them. One of his interests was a problem that had frustrated mathematicians for generations: the many-body problem, or more particularly, the three-body problem (determining the motion of three bodies under mutual gravitational attraction). Poincaré saw that it was much more difficult than others had realized. In fact, it appeared impossible to solve algebraically, so he turned to geometry. He introduced a space of several dimensions in which each state of the system at any time is represented by a point. This space is called phase space. It allowed him to turn numbers into pictures. Within this space he could look simultaneously at all possible behaviors of a system. This was a radically different approach, but it gave considerable insight. Although it wasn't a solution in the usual sense it gave Poincaré an indication of the complexity of the problem. The complexity, however, eventually discouraged him, and he gave up and went on to other things.

The problem was so difficult, in fact, that few advances were made for many years after Poincaré's work. The next important advance didn't come until the 1960s when the American mathematician Stephen Smale looked at dynamical systems (more specifically, the three-body problem) from a new point of view and showed that it could be understood in terms of a stretching and folding in phase space, much in the same way a baker stretches and folds his dough when making bread.

About this time computers began coming on the market. They were crude and slow compared to modern computers, but they were a Godsend for anyone working on problems where a lot of simple calculations were needed. Edward Lorenz of Massachusetts Institute of Technology (MIT) used one of these early computers to model the weather, and what he found surprised him. He discovered that extremely small changes in the initial conditions had a significant effect on the weather. We now refer to this as sensitive dependence on initial conditions, and, as we will see, it is a major characteristic of chaos. Lorenz was unsure why weather was so sensitive but he realized his discovery would make long-range weather forecasting impossible and it was one of the key breakthroughs in chaos theory.

Chaos can take many forms. The turbulence you see in a fast flowing river, for example, is one form. A question of importance to many theorists in the early part of this century was: How does turbulence begin? Where does it come from? The Russian mathematical physicist Lev Landau addressed the problem in 1944 and presented a theory that he believed explained it. A similar theory was proposed in 1948 by Eberhard Hopf, and for years the Landau–Hopf theory was the accepted theory of turbulence. In 1970, however, David Ruelle, a Belgian mathematician working in Paris with Dutch mathematician Floris Takens looked into the Landau–Hopf theory and showed that it was lacking. Introducing a new entity into phase space—what they called a "strange attractor"—they were able to explain turbulence in a much simpler way. Harry Swinney of the University

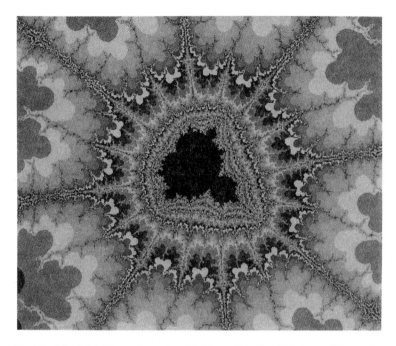

The Mandelbrot Set. The most complex object in mathematics. This is one of the smaller snowmen within the overall pattern. (George Irwin)

of Texas and Jerry Gollub of Haverford College verified Ruelle and Taken's theory in 1977.

But chaos isn't restricted to physical systems. Robert May, who was working at the Institute for Advanced Study at Princeton, showed that it also arises in biological problems. It was well-known that biological populations fluctuated wildly, depending on the food supply. An equation called the logistic equation had been developed to model these populations. May used this equation to show that the population could become chaotic.

Mitchell Feigenbaum of the Los Alamos National Laboratory followed up on May's discovery. Working with a hand-held

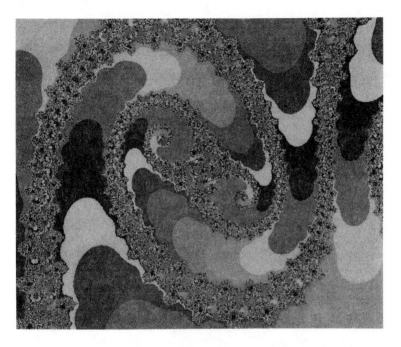

Closeup view of part of the Mandelbrot Set. (George Irwin)

computer, he was able to show that chaos wasn't as random as everyone had thought. The route to chaos was universal, and had a universal constant associated with it.

A few years later an interesting link was forged with another developing science—fractals. Fractals are structures that are similar on all scales. In other words, as you look closer and closer at them they always appear the same. Benoit Mandelbrot of MIT is generally acknowledged as the father of fractal theory; he gave us the name and many of the major ideas of the theory. Furthermore, using a computer, he generated one of the most puzzling fractals ever seen, an object that is referred to as the most complex object in mathematics. We refer

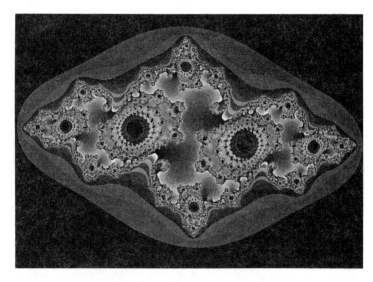

One of the Julia Set. (George Irwin)

to it as the Mandelbrot set. With a simple equation he was able to generate exceedingly beautiful and complex pictures, several of which are shown in this book. Fractals seemed to have no relation to chaos at first, but eventually they were shown to be closely related.

Astronomers took to chaos theory slowly. Poincaré, of course, developed some of the original ideas, but for years no one tried to do anything further with his problem. In the late 1970s, however, Jack Wisdom, who was then at the University of California, became interested in the gaps in the asteroid belt. Were they generated by chaos? It seemed possible. Wisdom developed a technique for dealing with the problem and showed that at least one, and perhaps all, the gaps were due to chaos.

Wisdom and others began to look at other objects in the solar system. Hyperion, one of the moons of Saturn, was an obvious candidate. Photos from Voyager indicated it was irregular

Chaos in the solution of equations. (George Irwin)

and was tumbling erratically. Could its motion be chaotic? Wisdom looked into this.

Another candidate was the planets. They certainly didn't look chaotic; they all appeared to be orbiting in a predictable way, and as far as we knew they had been doing so for many years. But was it possible that their orbits would be chaotic if you could follow them far enough into the future? Wisdom, along with Gerald Sussman built a specially designed computer that allowed them to follow the orbits of the outer planets hundreds of millions of years into the future (and past) and they got some fascinating results.

While Wisdom and Sussman were looking at the outer planets, Jacques Laskar in France was looking at the orbits of the

Planets and moons. Jupiter is in the foreground. (NASA)

inner planets, and he also obtained some interesting results. Others soon joined in the search.

But if there is chaos in the solar system, what about the stars? Pulsating stars are obviously good candidates. Is it possible that the pulsations can get out of control and become chaotic? This has been looked into.

Stars in binary systems, where one star is pulling matter from the other, are good candidates. One of the best examples of a system of this type is Her X-1, an x-ray source that is known to be associated with a neutron star. Several people have looked for chaos in this system. The bizarre system, Cyg X-1, which appears to contain a black hole, has also been looked at.

Stars orbiting the cores of galaxies are another candidate. Just as planets and asteroids in our solar system can become chaotic, so too, can stars. Considerable interest has recently centered around barred galaxies—galaxies with a bar–like structure through their center.

It was shown early on that chaos is associated with a particularly difficult type of mathematical equation called a nonlinear equation. It was therefore reasonable to look at one of the most famous of these equations—Einstein's equation. Considerable effort is now underway to study the effects of nonlinearity in general relativity, and chaos seems to play an important role. Chaos has also been shown to be associated with black holes. Objects orbiting two black holes, for example, can be chaotic under certain circumstances.

Einstein's equations are also the basic equations of cosmology and the early universe. A model of particular interest in relation to the early universe is called the Mixmaster model, named for its mixer-like oscillations (similar to the oscillations that occur in the mixer in your kitchen). Such oscillations may have occurred early in the formation of the universe, and may be chaotic.

And there's the theory of atoms and molecules—quantum theory. It also plays an important role in astronomy. We know that chaos occurs in the realm of classical mechanics. Isn't it

A field of stars. Some of them may be chaotic. (Lick Observatory, University of California, Santa Cruz, Calif. 95064)

possible it could also occur in the atomic realm described by quantum mechanics? We will look into this in one of the latter chapters of the book.

Finally, one of the great aims of humans over the past few decades has been a theory of everything—a theory that will explain all of nature. What effects does chaos have on our ability to formulate such a theory? We will look into this.

Before we can look at what effect chaos has in astronomy we have to learn something about the theory itself. We will begin that in the next chapter.

2

The Clockwork Universe

*O*ver many years a procedure for determining the position and velocity of particles was formulated, a procedure we now call classical mechanics. According to this theory, the future of any particle could be determined from its present position and velocity, once the forces acting on it were known. Classical systems behaved in a regular organized way, and their future could be predicted through use of the appropriate mathematical equations. The world, it seemed, was deterministic. In this chapter we will look at the rise of this determinism.

COPERNICUS, KEPLER, AND GALILEO

For the first thousand years in the history of humankind the Earth was dominant. Everything in the heavens revolved around it, everything was subject to it. It was the center according to biblical teachings, and according to the universal model that was developed during this period—the Ptolemaic model.

But when astronomers looked into the heavens they saw things that were difficult to explain with an Earth-centered system. Throughout most of the year the planets moved relative to the stars, but occasionally they stopped, changed direction for a while, then stopped again and resumed their forward motion. This retrograde motion, as it was called, was confusing, and the only way astronomers could explain it was by introducing tiny circular orbits called epicycles; these epicycles were superimposed on the larger orbit that was centered on the Earth.

It was a complex model, and to make things worse astronomers eventually found that simple epicycles weren't enough; they needed epicycles upon epicycles to fit observations well. Was it an accurate representation of nature? Ptolemy wasn't sure, but it could predict the motion of the planets, and eclipses, far into the future.

For over a thousand years the Ptolemaic system stood unchallenged. Then came Nicolas Copernicus. Although he admired the ingenuity of the system, he felt uncomfortable with its complexity. Was such a complicated model really needed? Nature, it seemed, would take the simplest and most economical route, and the Ptolemaic system was far from simple. Perhaps the universe appeared so complex because we were looking at it wrong.

Born in north Poland in 1473, Copernicus received his religious training at the University of Cracow and at the Universities of Bologna and Padua. Along with this training he studied mathematics and astronomy.

Copernicus was soon attracted to the works of the early Greek philosopher, Ptolemy. After considerable study he came to the conclusion that some of the difficulties of the Ptolemaic system could be overcome if the sun was placed at the center, rather than the Earth. Retrograde motion, in particular, could be explained without the use of epicycles as long as the inner planets traveled faster in their orbits than the outer ones. We see the phenomenon here on Earth when we are in a train on a curved track, passing another train. As we speed by, the neighboring train appears to move backwards, but once we are well past, we see that it is actually moving in the same direction we are.

Copernicus.

In a heliocentric, or sun-centered system, the same thing would happen as the Earth passed a planet such as Mars. For a while it would appear to move backward, but eventually it would resume its forward motion.

When Copernicus worked out the details of his model, he saw that it was simpler and more elegant than the Ptolemaic system, but strangely, he wasn't able to eliminate epicycles, and furthermore, he wasn't able to predict the positions of the planets any better than the older model. Still, he was convinced that it was better.

Copernicus was relatively young when he began working on his model, probably not over 40. He knew it wouldn't be looked upon favorably by the Church; he had dethroned the Earth, making it no more important than the other planets. The sun now held center stage. But according to Church doctrine the Earth was central; everything in the universe was secondary to it. Copernicus therefore kept his ideas to himself, but as the years passed he became more and more convinced of the validity of the system and eventually began circulating pamphlets to some of his friends describing it. One of his friends, a mathematician named Joachim Reticus, became fascinated with the model and encouraged Copernicus to publish. Knowing the consequences, however, Copernicus held back until he was about 70, and had only a short time to live. The book came out in 1543. Only a few hundred copies were printed. About 200 still exist in various libraries around the world. Two significant deviations from Copernicus' original manuscript were made. First, the title was changed from *De Revolutionibus* (On the Revolutions) to *De Revolutionibus Orbium Coelestium* (On the Revolutions of the Heavenly Spheres). This was no doubt done to take the emphasis away from the Earth. Second, a preface was added stating that the system was only a model and not necessarily a true representation of the universe. It was written by a friend of Rheticus.

Copernicus only saw one copy of the book, and it was brought to him on his death bed. The book had little impact at first. Epicycles were still required and the predictions did not agree with observations any better than they did for the Ptolemaic system. Little by little, however, its impact began to be felt.

One of those attracted to the new system was the German astronomer, Johannes Kepler. Born in Weil in 1571, Kepler was a man of many faces. He made some of the most important discoveries in the history of science which, according to his writings, took him to the heights of ecstacy. His personal life, however, was filled with unhappiness and tragedy. He had little self-confidence, was neurotic and sickly, and at one point even referred to himself as "doglike." Yet he was intelligent, talented, and a first-rate mathematician.

Early on Kepler became convinced that there was a divine, but precise, mechanism behind the motion of the planets. He

Kepler.

was sure the details of this mechanism were within his grasp, and once understood he would be able to predict the positions of the planets in their orbits for all time. But to gain access to this understanding he would need observations, preferably years of positional data, and only one person in the world had such data—the Danish astronomer Tycho Brahe, who was now at Benatek Castle, near Prague.

Like Copernicus, Kepler studied for a position in the Church, but his teachers soon saw that he was not suited for the clergy. His tremendous mathematical talent was evident even then, however, and he was encouraged to go into teaching. Kepler followed their advice, but soon regretted his decision. He had little control over his classes and eventually became disillusioned and disheartened; furthermore, his interest in the planets was now taking up much of his time, and he began looking upon his teaching duties as an intrusion.

The religious turmoil around him was also putting him in danger. He knew he had to leave. One day he got a letter from Tycho. Tycho had read one of his books and was so impressed he offered him a job. Kepler was overjoyed. By now he had become completely discouraged with teaching. Working for Tycho would enable him to direct all his time to astronomy. Furthermore, he would have access to the tremendous data Tycho had accumulated.

In February 1600, he packed up his family and his belongings and headed for Benatek Castle. The ride must have been difficult, but inside Kepler was bursting with excitement; he was sure it was a turning point in his life. Tycho, as it turned out, was also particularly eager to meet Kepler. He had been in a feud with another astronomer over his "world system," and he hoped Kepler would be able to help him. Tycho's system was strange. His was different from both Copernicus' and Ptolemy's—a combination of the two with the sun at the center. The planets orbited the sun, but the sun revolved around the Earth. Tycho needed Kepler, but he was cautious; from reading his books he knew that Kepler had the power to do things that he couldn't, and might overshadow him. He was unsure how to deal with him.

Kepler's expectations were high when he reached the castle, but within days his balloon of enthusiasm had burst. The two men were hopelessly different and Kepler found it was almost impossible to deal with Tycho. Arrogant and overbearing, Tycho was surrounded day after day by students, in-laws, and hangers-on, and to his dismay, Kepler soon found that Tycho had no intentions of allowing him access to his data. He doled only a small amount out at a time. The two men quarreled continuously, with Kepler threatening to leave several times. He actually packed his bags and ordered a stagecoach once, but Tycho finally persuaded him to stay.

Kepler was determined, however. The motions of the planets were not a problem he was going to abandon easily. He was confident he could solve it once he had access to Tycho's data, and within a year he had his opportunity. Tycho died after drinking too much beer at a royal dinner party, and Kepler inherited everything—his data and his position.

"Let me not have lived in vain," Tycho pleaded on his deathbed. He urged Kepler to use the data he had accumulated to vindicate his world system. It is not known what Kepler's reply was, but there is little doubt that by this time he was a confirmed Copernican.

Within a short time Kepler began his "War with Mars," as he called it. He thought at first that he would be able to determine its orbit within a week—eight days at the most. Eight years later he was still battling with the planet.

He wrote about his discoveries in two books. Considerable insight into his personality can be gained from them. He writes, not only about his discoveries, but also about his blunders and frustrations. He is, in fact, quite candid about how stupid he was at times, and how he overlooked things that should have been obvious, but buried within this chitchat are three gems, now known as Kepler's three laws of planetary motion.

Kepler's first law states that the planets orbit, not in circles, but in ellipses (egg-shaped curves). Kepler claims to have used 70 different circles in trying to fit Mars' orbit before giving up. Finally he tried an ellipse, with the sun at one foci. It was the answer.

Kepler's talents were particularly evident in his second law. It took considerable ingenuity and insight to arrive at such an important result. He discovered that the planets do not travel around their orbits at a constant rate, rather the line joining the sun and the planet (called the radius vector) sweeps out equal areas in equal times.

One of the consequences of this law is that when a planet is near perihelion (its closest point to the sun) it travels fastest, and when it is near aphelion (its farthest point from the sun) it travels slowest. In between, it travels at intermediate speeds.

It is easy to see now why Copernicus had problems. Although he had the right idea in placing the sun at the center of

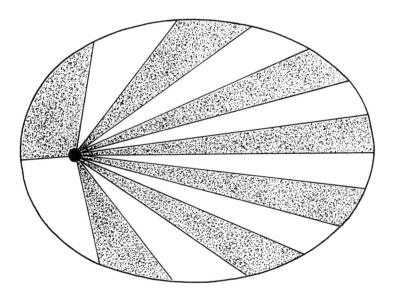

An illustration of Kepler's second law. Equal areas are traced out in equal times. This means the planets travel fastest when near perihelion.

the solar system, he retained circular orbits. Furthermore, he had the planets going around at constant rates. This is what forced him to retain epicycles.

Kepler's third law is a little more mathematical than the others. It tells us that the squares of the orbital period of the planets (time to go around sun) are proportional to the cubes of their average distance from the sun. As we will see later, this law played a key role in Newton's verification of the law of gravity.

Kepler's laws are still an integral part of astronomy, used not only in the study of planetary motions, but also in the study of binary systems, neutron stars, black holes, and the rotation of whole systems of stars—galaxies. Kepler was the first to introduce order and structure into planetary dynamics, a tremendous achievement considering that most of the basic concepts of physics had not yet been discovered. He played a major role in paving the way for the establishment of the Copernican system.

Kepler led an event-filled, yet tragic life, and his trail of tragedies followed him to the grave. He lost his position at Benatek Castle, and while traveling on horseback to Ratisbon to collect wages owed to him he fell ill and died. He was 48.

The telescope was invented during Kepler's lifetime, but he had little opportunity to use it. He wrote to Galileo hoping to get one, but Galileo, fearing competition, replied that he was too busy to make him one.

Born in Pisa, Italy in 1564, Galileo started out in medicine but soon became bored. At that time the church did not allow anyone to cut into the human body, so little was known about anatomy. He finally dropped out of school and studied on his

Galileo.

own for a few years. During this time he built many ingenious mechanical devices. His talents eventually came to the attention of a local nobleman who got him a position at the University of Pisa. For several years he lectured in mathematics and astronomy, and was popular with the students. Eventually, though, he made so many enemies amongst the faculty that he was let go, but he soon got a similar position at the larger and more prestigious University of Padua and he remained there for 18 years.

By the time he was 45 he was a respected scientist, a popular teacher and a friend of nobility, but for Galileo this wasn't enough. He wanted more, and an opportunity came one day while he was visiting Venice. He heard of an optical instrument that could magnify distant objects. Hurrying back to Padua he constructed a three power telescope, and over the next few years he increased its power to 30. It was the turning point of his life.

Although he didn't invent the telescope, Galileo made it into a powerful scientific instrument, and of particular importance, he turned it toward the heavens and unveiled many of its mysteries. He made discovery after discovery. He saw that the moon was cratered, he saw that Venus presented phases, he saw the moons of Jupiter, and he noticed that the sky contained many stars that could not be seen with the naked eye. He published an account of these discoveries in a book titled *The Starry Messenger*, and almost overnight it made him a celebrity.

By this time Galileo was a confirmed Copernican. Fully familiar with the Church's opposition to his cherished system, Galileo now saw the opportunity that he had been waiting for. With the telescope he could prove, once and for all, the validity of the Copernican system. Furthermore, he was sure his fame would help.

Still, he would have to wait for the right moment, and it came in 1623 when his friend Cardinal Barberini was elected Pope. Barberini was educated, and had an interest in science. Galileo rushed to present his case, sure that Barberini would be easily persuaded, but Barberini was cautious, and far from convinced by Galileo's arguments; he told him to wait until there was more proof.

Galileo was heartened by the Pope's interest, but instead of searching for more proof he began writing a book. Over a period of four years he shaped and refined it, publishing it finally under the title *Dialogue on the Two Chief Systems of the World*. The book took the form of a discussion between three philosophers. Salviato, the most intelligent of the three, was the mouthpiece of Galileo; Sagredo, also intelligent, was quickly convinced by Salviato's arguments; and Simplicio, stubborn and simple as his name implies, used the same arguments that Barberini used earlier with Galileo.

When the Pope read the book he was shocked. He quickly ordered publication stopped and he seized all unsold copies. Galileo was summoned to Rome in 1633, and was brought before the Inquisition. Although he wasn't tortured he was threatened, and in the end he pleaded guilty and signed a document denying his views on the Copernican system. He was placed under house arrest, and remained there until he died in 1642.

Galileo vowed never to speak or write about the Copernican system again, and indeed he had little opportunity. In the final eight years of life, while under house arrest, he did, however, manage to write his greatest book, *Dialogue Concerning Two New Sciences*. In it he outlined his views on nature and inertia. Inertia, as he defined it, was the tendency of a mass to resist change in motion, and it is an important concept in physics today. Galileo also showed that, neglecting air, all objects fall at the same rate, and a cannon ball travels farthest if projected at an angle of 45 degrees. His most important contribution, however, was the introduction of the scientific method: He initiated the idea of setting up and performing experiments.

Still, the underlying laws of nature had not been discovered, and while mathematics had been employed by Kepler and others, it was not yet an integral part of science. But the time was ripe, and in the same year that Galileo died, Isaac Newton was born in Woolsthorpe, England. He would bring mathematics into science, and with it science would be changed forever.

NEWTON AND THE INTRODUCTION OF MATHEMATICS INTO SCIENCE

Newton was an only child, born prematurely to a widow on a farm in Woolsthorpe. His mother remarried shortly after he was born, and he was raised by his grandparents, who lived only a mile and a half down the road. He was solitary and liked to play by himself, keeping busy by building windmills, simple clocks, and sundials. His ingenuity was evident even then, but few noticed. At school he was inattentive and tended to daydream, yet even when he fell behind he could easily cram everything in his

Newton.

head in the last couple of days before the exam, and place at the top of the class. This made him unpopular with his classmates.

Newton's mother eventually brought him back to her farm, hoping to make a farmer out of him, but it was soon evident he was not suited for farming. Sent to fetch the cows he would take a book with him, and hours later would be found sitting in the shade of a tree reading, having completely forgotten about the cows. But he was an excellent student, and his teachers encouraged his mother to let him go to university.

At Trinity College, Cambridge, he took the usual undergraduate curriculum, and although he did well, there is little evidence that he was outstanding. His talents did, however, come to the attention of one of his teachers—Isaac Barrow.

Newton received his bachelor's degree in 1665, the same year that the plague struck England, and because of the plague he was forced back to the farm at Woolsthorpe for two years. He appeared to do little during this time, sitting in his study, sometimes wandering in the garden, but the two years he spent here are two of the most important in the history of science. Newton was at the peak of his intellectual powers and he made full use of them. After watching an apple fall from a tree one day he made one of the most important discoveries ever made in science. He formulated a law of gravity: *All objects in the universe attract one another with a force that is inversely proportional to the squares of their separation.* It was strange language to most, and difficult to comprehend, but it brought new order to the universe.

The discovery of this one law would have engraved his name in the Halls of Science for all time, but he did not stop with it. During his two years at Woolsthorpe he also discovered the basic laws of motion, many of the properties of light and lenses, and he invented calculus. Quite a feat for a 25-year-old.

While the law of gravity was important in itself, and would soon allow scientists to make predictions about the objects in the solar system, its real significance was that it was a universal law. It applied not only to objects on the earth, but to the entire universe.

Did Newton publish this incredible discovery? Surprisingly, he was satisfied merely to have discovered it. He put his calculation away and hardly thought about it again for over a decade.

When the plague was over, Newton returned to Trinity College and two years later was named Lucasian Professor of Mathematics when his teacher Isaac Barrow stepped down. Over the next few years Newton's major preoccupation was light and optics. He experimented with lenses, carefully analyzing light beams as they were refracted and reflected. Upon grinding a prism he found that a beam of white light could be broken up into the colors of the rainbow. This convinced him that white light was composed of all colors. He also invented the reflecting telescope and upon presenting it to the Royal Society, was elected to its membership. Several of the members encouraged him to present a paper summarizing his discoveries in optics, and Newton accepted. It was a mistake that became etched in his memory, an experience that left a foul taste in his mouth. To his dismay several of the members argued vehemently and aggressively against his ideas. He answered their objections and questions patiently at first, but as they continued he became frustrated and eventually vowed never to subject himself to such harassment again. He decided to keep his discoveries to himself.

Locked away from the world, as a result, were his important discoveries on the dynamics of motion and the law of gravity, and they may have stayed locked away if it hadn't been for Edmond Halley. In January 1784, Halley, who was then only 27, was talking to Christopher Wren, part-time scientist and architect of St. Paul's Cathedral, and Robert Hooke, who had made a number of discoveries in physics and astronomy. Both men were considerably older than he. They were discussing the planets, wondering why they orbited the sun the way they did. Hooke and Halley were convinced the force between gravitating bodies fell off as the inverse square, but if this was so, what type of orbit would they trace out? Furthermore, how did Kepler's laws fit in? Could they be derived from this force?

Halley.

Hooke proudly boasted that he had solved the problem, but wasn't ready to make it public. Wren doubted him and offered a prize of a 40 shilling book to either of them if they could produce the proof within two months.

Halley struggled with the problem, and when Hooke never came forth with his proof, Halley decided to visit Newton. He had visited with him earlier and got along well with him. Newton invited him into his study courteously, but by now he was beginning to view everybody with suspicion. They talked for a while, then Halley ask him the key question. What type of orbit would be produced by an inverse square force law? "An ellipse," answered Newton immediately. Halley was surprised by his lack of hesitation. Newton then offered to get the calculation for him, but he couldn't find it, so he promised to send it to him.

Five months later the calculation arrived; it was nine pages long. Halley was amazed. On those nine pages were mathematical equations that explained everything—the elliptical orbits and Kepler's laws. Convinced that Newton had to have much more stored in his trunks and closets, Halley rushed back to Cambridge, and soon found that his suspicions were correct. Newton had much more that he hadn't published. Halley encouraged him to make it public, but Newton remembered his earlier experience and was reluctant. Finally, however, he relented, and Halley presented the project to the Royal Society. They agreed that publication was important, but they didn't have any money. The last book they had published (a history of fishes) had been a financial disaster and the Society was close to broke. Who would pay for the publication? Newton, although fairly wealthy, refused, so it fell on Halley's shoulders, and he could ill-afford it. Nevertheless he paid for the publication, and for the next two years oversaw the project.

Newton had done little on dynamics for several years when Halley approached him, but his interest was quickly rekindled, and he threw himself wholeheartedly into the project. His man-servant said he hardly ate or slept during the time he was working on it. But Newton wanted the book to be complete, and indeed, it was.

Titled *Mathematical Principles of Natural Philosophy*, or the *Principia* for short, it had three major parts. Newton's law of gravity was contained in the third part. Upon hearing of the project, Hooke argued strenuously that Newton should acknowledge him as discoverer of the inverse square law. Newton had many dealings with Hooke and still had a sour taste in his mouth from them. Hooke had been one of the people who had argued strongly against Newton's ideas on optics. On another occasion he had even ridiculed him when he made a slight mistake in a calculation, and Newton had never forgiven him. He said that he would prefer to withhold publication of the third part rather than have Hookes' name in it. Halley acted as mediator and finally smoothed things over. Hooke was not acknowledged.

Several people had, indeed, discussed the possibility of an inverse square law. Hooke was one, but he did nothing with the idea, whereas Newton incorporated it into his law of gravity and demonstrated that it was valid by applying it to the moon. Furthermore, he did this many years before Hooke and others even thought about it.

The *Principia* is now considered to be one of the most important scientific books ever published. Despite being difficult to read, it sold well, and within a short time established Newton as one of the greatest scientists who ever lived. It was, without a doubt, a turning point in the history of science, and marked the beginning of theoretical physics. Newton brought mathematics into physics and astronomy, and through it gave us a new understanding of nature.

Near the beginning of the book were Newton's three laws of motion. Galileo had had some of the basic ideas, but Newton was the first to state them in three simple laws. His first law stated that all bodies in straight line uniform motion continued in straight line uniform motion unless acted upon by a force. His second law told us that the body would accelerate if acted upon by a force, and his third, that for every action there is an equal and opposite reaction. We see his third law at work every day. When we hold a hose, for example, we feel a backward force on our hands as water spurts out in the forward direction.

The *Principia* was soon the bible of the new dynamics.

THE POWER OF PREDICTION

Using the law of gravitation, scientists could now calculate the orbits of the planets. It wasn't a simple technique, and could only be done by someone who had mastered the intricacies of the theory. Even Newton had difficulty applying his theory. In fact, he was not happy with the result he got when he applied it to the motion of the moon around the Earth. He had to fudge slightly to make things agree with observation (this was mostly due to the inaccuracy in the known distance and mass of the moon). He found it easy to deal with two bodies, but when he had to deal with more, the difficulty of the problem increased considerably. The Earth–moon problem was really a three-body problem because the sun strongly influenced the moon. Newton eventually developed a technique called perturbation theory for getting around this problem. He would solve the relevant two-body problem, then add in the contributions of the third body (and other bodies). The technique worked well as long as the perturbations from the other bodies were small.

One of the first to use Newton's law of gravitation was Halley; he predicted the return of a particularly bright comet that was seen at the time. Using Newton's theory, he calculated the orbit of the comet and predicted it would return in 1758; he even showed where in the sky it would appear. At first there was little interest. Halley's comet, as it is now called, has a period of 76 years, and its reappearance was far in the future. But as the 1750s approached interest resurfaced. Halley had died in 1742, but many people had now become obsessed with being the first to spot the comet. If Halley was correct it would be a tremendous verification of the power of Newton's equations.

In late December—on Christmas eve—in 1758 a German farmer, Johann Palctzsch, saw the comet almost exactly where Halley had predicted it would appear. The world was galvanized; there was a clockwork mechanism behind the workings of the universe after all, and Newton had led the way in understanding it.

Another verification came 42 years later. On New Year's eve 1800, Sicilian astronomer Giuseppe Piazzi, while making a map of the sky, came upon an object that was not on any of the earlier maps. He thought it might be a comet, but it wasn't fuzzy like most comets; it had a distinct disk. He reported the object to Johann Bode at the Berlin Observatory, and Bode suggested it might be a planet. Several years earlier Johann Titius had shown that there was a strange numerical relationship between the orbits of the planets (to each of the numbers 3, 6, 12, 24, ... add 4 and divide by 10. The resulting sequence is the distances to each of the planets in astronomical units.). The problem with the Titius sequence was that it predicted a planet between Mars and Jupiter and none had been found. Bode was sure this was the lost planet.

The first step was to calculate its orbit. But Piazzi had obtained only a few closely spaced observations before it disappeared into the day sky, and with the techniques available at the time it was impossible to determine the orbit. Several points with considerable spacing were needed. The brilliant mathematician Karl Gauss heard about the problem and was delighted, as he had just devised a powerful new technique for determining orbits in which only a few closely spaced observations were needed. He could test it. Calculating the orbit, he predicted where the object would reappear, and it reappeared on schedule. Astronomers soon found to their disappointment, though, that it wasn't the long lost planet between Mars and Jupiter; it was too small. It was an asteroid.

Another opportunity to test Newton's law in another way came a few years later. In March 1781, William Herschel had discovered an object in the sky that was not on any of the maps. He thought at first that it was a comet, but it presented a sharp disk, and moved much too slowly. Herschel reported the object to the Royal Society. With further observations it became clear that it was a planet. Within a couple of years Simon Laplace in France, and others, had calculated its orbit. It was soon evident, however, that something was wrong. Within a decade the planet was deviating from its predicted orbit, and within a few decades it had deviated significantly. Corrections were made for the presence of Jupiter and Saturn, but they didn't help. A number of people began to suspect Newton's law. Did it apply to distant objects such as this one? Laplace and others suggested that there was another planet beyond Uranus that was perturbing it, pulling it out of its predicted orbit.

The mystery persisted for another 30 years. Then in 1841 a 22-year-old student at Cambridge University, John Couch Adams, became interested in the problem. He was soon convinced that there was, indeed, a planet beyond Uranus that was perturbing it. He graduated a year later, took a position at St. Johns College, and began spending all his spare time on the problem. It was not an easy task; no one had ever tackled anything like it before. Adams would have to guess roughly where the new planet was, then apply a series of approximations to narrow in on its true position. In the early stages of the problem he was able to use circular orbits, and this helped considerably, but as he narrowed in on the planet he had to switch to elliptical orbits and things got much more complicated.

By September 1845 he felt he had determined the position of the new planet accurately enough for a search to begin. He took his results to James Challis, the director of the Cambridge Observatory, and to George Airy, the Astronomer Royal. Both men had encouraged him earlier, even though neither had much faith in mathematical techniques. They found it difficult to believe that anyone could predict the position of a planet using mathematics. So when Adams gave them his results they ignored him, sure that such a prediction wasn't important enough to disturb the observing schedule at the observatory. Adams persisted, but little was done, and he eventually became discouraged. In the meantime he continued to refine his calculations.

Unknown to Adams, a 34-year-old mathematician in France was taking up the problem about this time. Urbain Le Verrier had already established himself as an able mathematician. Within months he had a first estimate of the position of the perturbing planet. He presented his result to the Paris Academy of Science in late 1845, and although there was some interest, no one offered to look for it. Le Verrier pressed the matter, presenting a second paper a few months later, but still, there was no action.

Le Verrier's prediction made its way to Airy in England, and to Airy's surprise it agreed extremely well with Adam's prediction. Airy finally decided to take action, and instructed Challis to began a search, but instead of looking exactly where Adams had predicted the planet would be, Challis began a routine search of a wide area around it.

Le Verrier was getting impatient. Earlier, a young graduate student, Johann Galle, who was now at the Berlin Observatory, had sent him a copy of his thesis. Le Verrier had not replied at the time, but realized now that it might be the opening he needed. He sent a letter to Galle praising his thesis, and asking him if he would be interested in searching for a new planet; he enclosed its coordinates. Galle was flattered to be asked and quickly went to the director, asking his permission to make the search. As it happened there was some free time on the telescope, and on September 23, 1846, Galle and an assistant began the search. Luckily they had a new map of the region where the suspected planet was, and within an hour they found an object that was not on the maps. It was hard to contain their excitement, but they had to verify that it was, indeed, a planet. It had a sharp disk, but would it move among the stars? They checked the next night, and indeed, it had moved. It was a planet.

The planet was within a degree of Adam's and Le Verrier's prediction, and it was another triumph for Newton's theory. The news spread rapidly. Airy was shocked and to some degree embarrassed; he had delayed too long and had been scooped by the Berlin Observatory. He spent many years trying to live down the embarrassment.

Interestingly, Challis had actually seen the planet twice during his search, but hadn't recognized it as a planet. In fact, the planet had been seen dozens of times in the preceding years. Even Galileo saw it pass near Jupiter as he was studying Jupiter's moons.

As might be expected there was an uproar in Germany after Airy announced that Adams had predicted the position before Le Verrier. For years there had been an intense scientific rivalry between England and the continent, and officials in Germany were outraged that England was also claiming credit. Things were eventually smoothed over, however, and Adams and Le Verrier were both credited with having predicted the position of the new planet.

THE RISE OF DETERMINISM

As we saw earlier, one of Newton's early achievements was the invention of calculus. He discovered and developed the new mathematics soon after returning to Cambridge after the plague. He showed it to Barrow, but as was so often the case, he didn't publish. Strangely, though, he didn't make use of calculus in the *Principia*. Everything in the book was done geometrically, and years after it was published scientists were still using geometrical methods. Slowly but surely, however, mathematical equations began replacing the more cumbersome geometrical methods and soon mechanics had a strong analytical foundation. Scientists showed that mathematical equations could be written down for many natural phenomenon, and when these equations were solved, the past and future of the phenomenon were known.

Many of the early contributions to this transformation were made by the mathematician Leonhard Euler. Born in Basil, Switzerland in 1707, Euler attended the University of Basil. In 1727 he was invited to join the St. Petersburg Academy which was being formed in Russia. His interests, while in Russia, extended beyond mathematics; he began studying the sun and in 1735 lost the sight of one of his eyes while observing it. His output was nevertheless impressive, covering many fields of mathematics and physics. Within a few years, however, political turmoil developed in Russia and Euler went to the Berlin Academy of Science. He remained there until 1766 when Catherine the Great came to power in Russia. The St. Petersburg Academy had floundered during the turmoil, and she was determined to return it to its former prestige. She invited Euler back, who was by then considered to be the greatest mathematician in the world. In the same year that he returned he lost the sight of his other eye, and was totally blind, but this did not affect his output. His memory was legendary and he could manipulate figures in his head with ease—almost as if he had a blackboard before him.

During his life Euler published over 800 papers and books, making him the most productive mathematician in Europe at the time. After his death it took years to sort out the contributions he hadn't even bothered to publish.

Euler made contributions to all branches of mathematics, but one of his most important contributions was putting calculus on a strong foundation, in short, developing a theoretical basis for it. Equally important, however, were his applications of these methods to the motion of natural systems. He published several textbooks, some the first of their kind. In 1736 he wrote the first textbook in which Newton's dynamics of a point mass was developed analytically (without the use of diagrams). In 1765 he extended this analysis to solid objects in a second text. He also published some of the first texts in calculus, starting in 1755 with a text on differential calculus, and followed, in 1768 through 1774, by three volumes on integral calculus. The theory of differential equations was developed in the latter volumes. Euler also had an intense interest in astronomy, and in 1774 published one of the first books on celestial mechanics.

While he was head of the Berlin Academy of Science, the work of a young mathematician, Joseph Louis Lagrange, came to his attention. Euler was so impressed with him that when he left Berlin for St. Petersburg in 1766 he recommended him as his replacement as head of the Berlin Academy.

Lagrange's father wanted him to go into law, but at school Lagrange read an essay by Edmond Halley on calculus and was enthralled. He began reading everything he could on the new mathematics, and soon decided to become a mathematician. Lagrange continued Euler's program of using analytical methods in mechanics. One of his more important contributions was the development of generalized coordinates. Until then, each problem had to be set up and solved in a particular coordinate system (e.g., polar coordinates or cylindrical coordinates) depending on the nature of the problem. Lagrange introduced coordinates that applied to all systems, and with them he was able to write down an equation that could be applied to all problems in mechanics. He summarized his results in his book *Analytical Mechanics* which was published in 1758. Unlike most earlier books on mechanics, this one did not have a single diagram in it.

As we saw earlier, Newton worked out the problem of two interacting bodies in motion in considerable detail, but had difficulty with three interacting bodies—called the three-body problem. Lagrange developed a procedure for dealing with the three-body problem.

Lagrange moved to Paris in 1789, but was not mathematically productive the last few decades of his life. He died in 1813.

The torch was passed to Pierre Simon Laplace. Born in 1749, Laplace came to Paris when he was 18 with a letter of introduction to the great mathematician Jean Le Rond d'Alembert. D'Alembert couldn't be bothered with seeing him, so Laplace sent him a manuscript on mechanics. D'Alembert was so impressed he quickly changed his mind, and obtained a position for him at the university.

Laplace spent so much time extending and developing Newton's theory that he was eventually referred to as the French Newton. He spent many years developing a theory of gravitation, publishing his results in a monumental work called *Celestial Mechanics* between 1799 and 1825. He developed the basic equation of gravitational theory, an equation that was later generalized by Simeon Poisson.

Slowly but surely all areas of physics and astronomy came under the influence of mathematics. Equations were developed for heat, light, gravitation, electrostatics, and hydrodynamics. All of these things could be understood and described in terms of differential equations. If the initial conditions were known, the future development of any system could be calculated and predicted. Mathematicians found many of the differential equations difficult to solve but this didn't quell their enthusiasm. In many cases if the position and velocity of an object could be measured at any given instant they could be determined forever. Thus began the era of determinism. Laplace was one of its most ardent supporters. He even went as far as boasting that if the position and velocity of every particle in the universe were known, he could predict its future for all time. The difficulties of such a venture were, of course, obvious; still, it seemed possible.

Mathematicians eventually found, however, that not only were some of the equations difficult to solve, but some of them were completely unsolvable. At first these "special cases" were ignored, but eventually they were examined in detail, and a significant change began to occur in mathematics and science.

3

First Inklings of Chaos

B *y the early 1800s determinism had become firmly entrenched. Given the initial positions and velocities of interacting objects* subject to forces in a dynamic system, you could, through the use of differential equations, determine their positions and velocities for all time. In practice the procedure worked wonderfully for one or two bodies. Drop a ball over a cliff, and you could predict to a fraction of a second when it would land if you knew the height of the cliff. And if the Earth and Sun were the only two objects in the solar system, you could determine the Earth's orbit to any desired degree of accuracy. But bring three bodies into the picture and the problem became a monster, a maze of mathematical equations so complex that even the most competent mathematicians balked at them. And beyond three bodies . . . that was unthinkable.

Yet most physical systems consisted of many bodies. There were ten major objects in the solar system and dozens of minor ones. In fact, a system of particular interest at the time was a gas. You may not think of a gas as a system, but indeed it is, and in the early 1800s quite a bit was known about gases. The basic gas laws, for example, had been discovered, and physicists were sure they could be derived from the properties of the components of the gas, namely its molecules.

But how could you deal with a gas? Even a thimbleful contained billions of molecules. To apply Newton's laws to each one, and sum, was unthinkable—literally beyond comprehension. Furthermore, there was still considerable controversy at that time as to whether a gas was even made up of molecules.

Still, there had to be a way. Gases seemed to behave predictably. There had to be an underlying theory that explained their properties. A deterministic approach was out of the question, but averages and probabilities for complex systems could be calculated. And as we will see they were the key.

THE RISE OF STATISTICS

Probability theory is at the foundation of the kinetic theory of gases, but as you might expect, it was invented for something entirely different. Gamblers in the 1600s, like gamblers today, wanted the odds in their favor. They were sure some strategies were better than others, and wondered how to select the best one. A well-known gambler of the time decided to write to the French physicist Blaise Pascal. He explained to him that he always lost money betting on certain combinations in the fall of three dice, and he wondered why. The question perked Pascal's curiosity and together with Pierre Fermat they looked into the problem, and in the process they came up with some of the fundamentals of probability theory. In particular they gave probability a formal definition. If an event can occur in only one of a number of equally likely ways, p of which are favorable, and q unfavorable, the probability of the event occurring is p divided by $p + q$.

This was only the tip of the iceberg, however, and for many years the theory lay dormant. Gamblers no doubt took advantage of some of the new insights, but no one else seemed to care. It was not until the early 1800s that Laplace, starting with the simple, disconnected concepts of Pascal and Fermat, built probability theory into a serious branch of mathematics. He published his discoveries in a book titled *Analytic Theory of Probabilities*.

With the discovery of the basic principles, a number of people began applying the ideas to complex systems, and as a result "statistical mechanics" was born. Scientists were soon able to relate the microscopic state of a gas—the actual positions and velocities of the molecules—to macroscopic properties that could be measured.

MAXWELL, BOLTZMANN, AND THE KINETIC THEORY OF GASES

The first theory of gases was advanced by Daniel Bernoulli. Born into an amazing family of mathematicians and physicists in 1700—his uncle, two brothers, a cousin and several nephews were all mathematicians or scientists—Bernoulli (although born in Holland) spent most of his life in Switzerland. Today he is mainly remembered for his discoveries in fluid dynamics; the Bernoulli principle (as the velocity of a fluid increases, its pressure decreases) is a basic principle of physics.

Bernoulli postulated that gases are made up of elastic particles, rushing around at tremendous speeds, colliding with one another and the walls of the container. Gas pressure, he said, is due to particles striking the walls of the container. Until then scientists had assumed that pressure was due to a repulsion within the gas, most likely a repulsion between the particles making up the gas.

Bernoulli's theory was a step in the right direction; he introduced probability into the problem but he didn't have the mathematical tools to set up a detailed theory, and over a century passed before any further serious work was done. The new advances, however, came in a flood, and were associated with two names: Ludwig Boltzmann of Austria and James Clerk Maxwell of Scotland.

Born in Vienna in 1844, Boltzmann was the son of a civil servant. He received his Ph.D. from the University of Vienna in 1866. It might be hard to believe now, but throughout much of Boltzmann's life an intensive battle was being fought over the existence of atoms. Boltzmann was in the thick of it. He not only argued strongly for atoms, he threw everything he had into the battle; his whole life centered around it. Boltzmann found it difficult to believe that others—even well-known scientists such as Ernst Mach and Wilhelm Ostwald—scoffed at the idea. His struggle was so intense that it eventually took its toll. In 1905, sure that the battle was lost, he went into depression and committed suicide; he was 62. Ironically, within a few years atoms were accepted universally.

Building on Bernoulli's work, Boltzmann laid a mathematical foundation for the kinetic theory of gases. But he did much more than that. Thermodynamics, the theory of the relation between heat and mechanical work, was an emerging science at the time, and Boltzmann linked many of the ideas he developed for kinetic theory to it. A few years earlier, in 1850, the German physicist Rudolf Clausius had introduced a concept he called entropy; it was the ratio of the heat contained in a system to its temperature. He postulated that entropy would always increase in any process taking place in a closed system.

Boltzmann showed that entropy was also a measure of the disorder of a system, and he wrote down a formula for it in terms of the probabilities of the various states of the system. This formula, one of the great achievements of his life, was engraved on his tombstone when he died.

While Boltzmann was working on kinetic theory, Maxwell, in England, was also working on the same theory, making the same discoveries. Born on an estate called Glenlair, near Edinburgh, Maxwell's talent for puzzles and mathematics was evident at an early age. Like Newton he loved to build clocks, water wheels, and other mechanical devices. At the dinner table he would sometimes get so involved in a simple experiment with light or sound that he would forget to eat.

Maxwell.

When he was eight his mother died, and a year later he was sent to the Edinburgh Academy to begin his education. By the time he was 15 he was solving complex mathematical problems, one of which won him the mathematics medal of the Academy.

From the Edinburgh Academy he went to the University of Edinburgh where as an undergraduate he read two papers before the Royal Society, a considerable feat for one so young. In 1850 he left for Cambridge and was soon preparing for the highly competitive mathematical exams called the Tripos. Considerable prestige went along with placing first. Just before the exams, however, he got sick, and it looked like he wouldn't be able to take them. But Maxwell was determined; he wrapped himself in a blanket, sneezed and coughed his way through the exams, and placed second. Interestingly, although the person who placed first became a well-known theoretical physicist, he never approached the greatness that Maxwell eventually achieved.

When he graduated, Maxwell took a position at the University of Aberdeen. Within a few years a substantial prize was offered for an explanation of the nature of Saturn's rings. Were they liquid, solid, or composed of discrete bodies? Maxwell was soon hard at work on the problem; after a concentrated effort he showed that they were composed of myriads of small particles, like tiny moons, each in its own individual orbit. He won the prize, and in the process gained a thorough understanding of particle dynamics. In 1860 he decided to apply this knowledge to Bernoulli's theory of gases.

One of the first problems he faced was finding an expression for the statistical distribution of velocities of the particles making up the gas. Following Bernoulli he considered the gas to be composed of particles moving in all directions, and at all velocities. He derived an expression for the velocities, finding that they varied considerably. Plotting them, he saw that the distribution took on the shape of a bell. We now refer to this as a normal distribution, and as we will see, much of statistics centers around it. Johann Gauss showed, for example, that observational errors in astronomy, when plotted, gave rise to a similar curve.

Toward the end of his life Maxwell, like Boltzmann, grew despondent, and seemed depressed. Unlike Boltzmann, though, the struggle over atoms never seriously affected him. His despondency stemmed from sickness; his mother had died of cancer, and in his late forties he was overcome by the same disease. He kept his condition to himself for as long as possible, but finally he could no longer walk, and he was taken back to Glenlair, where he died within two weeks.

Despite his short life, Maxwell left a tremendous legacy. Besides formulating the kinetic theory of gases independent of

Boltzmann, he put electromagnetic theory on a firm foundation, writing down the basic laws of electricity and magnetism as four simple equations.

Over many years and as a result of many independent advances, statistics became a basic tool of science. Interestingly, though, all the advances in statistics were not made in the physical sciences; some were made in the biological sciences and some in the social sciences. In the social sciences two names shine brightest: Lambert Adolphe Quetelet and Francis Galton.

Quetelet, a Belgian astronomer who studied under Laplace, eventually became director of the Brussels Observatory. Although statistics was becoming an integral part of astronomy at the time, Quetelet's discoveries were not made in his chosen field. As a side-avocation Quetelet began delving into the physical characteristics of people. He measured the chests of Scottish soldiers, the heights of French soldiers, and he plotted them, finding that they fit a bell-shaped curve similar to the one Maxwell would find later. People, he discovered, varied around an average in a particular way, regardless of what physical characteristic you measured.

Galton extended Quetelet's work to heredity. He became particularly interested in whether intelligence was inherited, and this led to an investigation of other inherited properties such as height, color of eyes, and so on. His interest in intelligence may have stemmed from his own obviously high intelligence. A child prodigy, he could read at three, and was studying Latin at four. He trained to become a physician but when he inherited a large estate about the time he graduated he gave it up and began travelling. He tried his hand at meteorology for a while but eventually gave it up too. But when he began applying statistics to heredity he stumbled onto a gold mine. There was considerable controversy at the time over whether intelligence was inherited or due to environment. Through his work on statistics he was able to show that heredity was definitely an important factor, and he went on to show it was important in relation to many other things: height, color of hair.

He found that literally all physical characteristics were determined in some way through heredity.

As in the physical sciences, statistics was soon playing an important role in the social sciences. In fact, many of the developments paralleled those in the physical sciences. Important advances in one of the sciences would quickly be taken over and adopted to problems in other sciences. Statistics became the new approach, taking its place alongside the deterministic approach of Newton. The deterministic approach worked well for simple systems but failed when applied to complex systems. The new statistical approach, however, worked well when applied to complex systems. The parameters that described the system in this case evolved with time, not determinately, but through probability. This led to uncertainties in making predictions, but it was, nevertheless, an effective approach. Both techniques—the statistical and deterministic—were effective, and each applied to different types of systems. Yet strangely, there was little connection between the two methods. Both gave answers to problems, and predictions could be made that were born out by observations. But there were problems. One of the major ones was: Is there a connection between the two methods? In addition, there were serious limitations in the regions of applicability of the two theories. The deterministic method, for example, appeared to fail when applied to a system of more than two interacting objects.

THE THREE-BODY PROBLEM

As we saw earlier, Newton had considerable trouble with the Earth–moon–Sun system, a particular case of the three-body problem. This problem, as we will see, played an important role in uncovering chaos and because of this it's worthwhile looking into it in more detail. The problem is to determine the motion of three bodies (usually considered to be point objects) attracting one another according to Newton's law of gravity. In the general case there are no restrictions on the masses, or the initial conditions. Mathematicians soon found, however, that the general case was extremely difficult, and they began directing their attention to restricted, or special cases, hoping they would lead the way to the solution of the general case. There are several variations of the restricted case: one particle might be very much less massive than the other two, or two could be much less massive than the third.

For the three-body problem, nine quantities (called integrals) are needed for a complete solution. Laplace solved the problem for a special case, but was unable to solve the general case. It was assumed at that time that all the integrals needed for a solution could be written down as algebraic expressions. It therefore came as a shock in 1892 when H. Bruns showed that this wasn't true. Even though they all existed in theory some of them could not be expressed in terms of known functions.

In the three-body problem you start with the positions and velocities of three objects and determine their positions and velocities at later times using differential equations (they give the time development of the system). In many such systems the objects return to a configuration they had earlier; if this happens, they are referred to as periodic. Periodic systems are quite common, and as a result mathematicians eventually became particularly interested in them.

Let's take a closer look at them. Consider three bodies; they can be three stars or three planets or whatever, as long as they gravitationally attract one another. Each of them, as you know, will move in an ellipse, and therefore the distance between them will change continuously. This means the gravitational force between them also changes continuously. In some cases, however, if you start with certain initial conditions the three objects will come back to their initial positions, with their initial velocities. They will depart from them again, but after the same number of orbits will come back again, and so on. This is called a periodic orbit.

Early mathematicians concentrated on periodic orbits in an effort to find solutions for cases that were not periodic. A standard technique was to look at systems with periodic solutions, then make small changes in the initial conditions and see how it affected the orbits. They expected that these small changes would only induce small changes in the orbits.

The standard technique for dealing with the three-body problem was, and still is, perturbation theory. In perturbation theory you start with a known solution, for example the solution of the two body problem, then add in the contribution of the third body, assuming it is small. For example, in the Earth–moon–Sun system, you could start with the two-body problem, the Earth and the moon, then add in the Sun's influence as a perturbation. Scientists soon found, however, that this didn't work in the Earth–moon–Sun system because the Sun's contribution was too large.

When perturbation theory is used, the solution of the problem is expressed in a series; a good example of such a series is $(1 + 1/2 + 1/4 + 1/8 + ...)$. In practice scientists usually restrict themselves to the first few terms of the series; in the above case we might select the first three terms, which would give $1 + 1/2 + 1/4 = 1\ 3/4$ (if all the terms are retained the series gives 2, so $1\ 3/4$ is a good approximation; if we take more and more terms we get closer and closer to 2). If the perturbation series is to give a valid answer, however, it has to converge. All series do not converge; if the succeeding numbers are large (e.g., $1 + 2 + 3 + 4 + ...$) the sum gets larger and larger, and we have divergence. Selecting the first few terms from such a series does not give anything close to the correct answer. One of the major problems in relation to the three-body problem, therefore, was proving that the series converged.

An important breakthrough in the problem came near the end of the nineteenth century when George Hill, an astronomer at the U.S. Naval Almanac Office, discovered a new method for attacking it. Until then everyone started with the bodies orbiting in perfect ellipses, then added in perturbations that modified the ellipses. Hill decided to start with the known solution for a special case; the orbit in this case was not a perfect ellipse. He then added the perturbation to this solution. It turned out to be a particularly useful technique, and most people that worked on the problem after him took advantage of it.

POINCARÉ, OSCAR II, AND CHAOS

The first glimpse into chaos came in an unexpected and strange way. In the late 1880s a contest was proposed to celebrate the 60th birthday of the King of Sweden, Oscar II. The prize, which consisted of 2500 crowns and a gold medal, was to be awarded on the King's birthday, January 21, 1889. Besides receiving considerable renumeration, the winner would receive recognition and prestige, so many people were eager to enter. Four problems were suggested by the eminent German mathematician Karl Weirstrass.

Three of the problems were in pure mathematics, and one in celestial mechanics. A few years earlier Weirstrass had heard of a problem related to the solar system. Julius Dedekind, another well-known mathematician, had hinted that he had shown mathematically that the solar system was stable. But he never wrote out the proof, and he died without letting anyone in on his secret (if indeed he did solve it). Weirstrass tried his hand at the problem, but with little success. He decided to include it as one of the four problems.

What do we mean by stable? The best way to understand it is to consider a block on a table. The downward force due to its weight passes through the surface supporting it and there is an equal and opposite reactive force that balances it. This state is stable. But take the block and turn it so it is balanced on one edge. The force due to its weight still passes through its point of support and it is in a balanced state. But is it stable? To determine this we need to look at nearby states. If we tilt the block slightly, the downward arrow no longer passes through the point of support and we no longer have a balanced situation. In fact, the block quickly tips to one side. In short, a

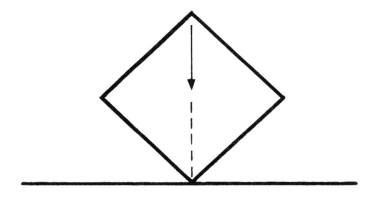

A slight tilt and the arrow no longer runs through the lowest point. This is an unstable situation.

slight change creates a large change in the system; the state is therefore unstable. If you do the same thing for a ball sitting on a table; in other words, if you push it slightly, it is easy to see that the two forces remain balanced, and a large change does not occur. This state is stable.

How does this apply to the solar system? What is important here is that a small change causes a large change in the system when it is unstable. If the solar system were unstable, weak forces, say those between the planets, could eventually cause the entire system to change dramatically. We know the planets have been going around the sun for billions of years in roughly the same way, so the system appears to be stable. Is it possible, however, that the orbits are slowly changing in such a way that the planets will eventually fly off to space? At the turn of the century this was considered to be a serious problem.

So the problem was cast out: Is the solar system stable? Actually, all that was specified was a solution in the form of a convergent series, which would imply stability. One who took it up was Henri Poincaré of the University of Paris. Born in Nancy, France in 1854, Poincaré is frequently referred to as the last universalist, in other words the last person to have a work-ing knowledge of all branches of mathematics. Indeed, not only was he at home in all branches of mathematics, he made im-portant contributions to most of them. Since his time no one has been able to do that, and it is unlikely anyone ever will again; mathematics has become too complex.

Poincaré's fame was so great, in fact, that just after World War I when the English philosopher Bertrand Russell was asked who was the greatest Frenchman of modern times he answered, "Poincaré," without hesitation. Thinking he was referring to Raymond, Henri's cousin who had become president of France, the questioner frowned. "No, not that Poincaré," Russell said quickly. "Henri Poincaré."

Although he was, without a doubt, one of the great French-men of his time, Poincaré showed little promise when he was young. He was brilliant in many ways, but slow in others. He read early, and voraciously. Furthermore he had a photographic memory, easily able to recall everything he read. But his physical development was slow; he had both poor coordination and eye-sight, and gave the impression that he might be slightly retarded.

His poor eyesight was, in one respect, a blessing. He couldn't see the blackboard well when he was young, and there-fore listened carefully, filing everything away in his mind. Before long he found that without taking a single note he could easily reproduce everything that was said—and add a few excellent suggestions for good measure.

He first became interested in mathematics when he was about 15, and by this time his ability to perform calculations in his head had become so acute he seldom worked anything out on paper. He preferred to do the entire calculation in his head, committing it to paper only after he had the final answer.

Poincaré took the exam for his bachelors degree when he was 17. Arriving late for the exam he got flustered on a problem related to convergence and almost failed the mathematics part. But he would never do this again. He placed first in mathematics in the entrance exams for the School of Forestry, and when he went to Ecole Polytechnic he continued to amaze his teachers with his mathematical talent.

In 1875 he left the Polytechnic to go to the School of Mines; he was going to become an engineer. The curriculum, however, left considerable time for him to work on mathematical problems, and it soon became obvious that his talents were being wasted in engineering. Three years later he was awarded a Ph.D. from the University of Paris for work he had initiated while at the School of Mines. One of his examiners, after looking at his thesis, said, "It contains enough material to supply several good theses."

In 1879 he went to the University of Caen as professor of mathematics, but his stay was brief. He was soon back at the University of Paris, and at 27 was one of their youngest professors. It was shortly after he arrived at the University of Paris that he heard of the prize to be awarded for the solution of a mathematics problem on Oscar II's birthday.

It was a challenge he couldn't pass up; winning the prize would give him recognition throughout Europe, and as he was just beginning his career, it was something he needed. He selected the problem on the stability of the solar system. As a first approximation it was a nine-body problem—eight planets and the sun (Pluto had not been discovered yet). In practice, however, the minor components of the solar system would produce perturbations on the planets so it was actually closer to a 50-body problem.

Poincaré saw immediately that he would have to apply some approximations; in other words, he would have to restrict the problem. Fifty, or even nine bodies, were far too many to consider seriously. He restricted himself to the three-body problem, and soon found that even it was extremely difficult in the general case.

Poincaré was familiar with the technique developed by Hill (starting with the known solution for a particular case), and he used it as a starting point. As we saw earlier, however, Bruns had shown that all the required integrals could not be obtained in simple algebraic form. Poincaré therefore decided to try a geo-metric approach. He started by plotting the three orbits in what is called phase space. Phase space is quite different from ordinary physical space, and for mathematicians and scientists it is, in many ways, more convenient, particularly when dealing with complex problems. In physical space the planet goes around the usual Keplerian elliptical orbit, and we make a plot of the planets position against time. As theoretical physics developed, however, Lagrange, Hamilton, and others found that position and velocity were more useful and convenient than position and time. Actually, instead of velocity they used momentum, which is velocity multiplied by mass. The new coordinates of the planet were therefore position (called q) and momentum (called p).

If we consider p and q as the new coordinates and plot them for each position of the planet we are in phase space. Again, we can look at the time dependence of the planet in phase space, and we get an orbit.

Dealing with three orbits in phase space was difficult, so Poincaré considered the restricted problem where one object was much less massive than the other two. It was in looking into this problem that Poincaré's genius really showed. Rather than examine the entire orbit he took a slice through it, a slice that we now refer to as a Poincaré section. Each time the planet passed through this slice it would make a mark, or at least we can assume it made a mark.

Poincaré considered the pattern of marks the planet made as it passed through again and again. Periodic orbits were easy to identify. After a certain sequence of points the planet would pass through the first point, then through the second, and so on. The whole pattern would be repeated.

The overall object was, of course, to prove that the system was stable, so what he was doing had to be related to stability. For stability, as we saw earlier, the series that was generated in the solution had to converge. Poincaré was sure that the pattern of points on his slice would tell him something about convergence. He hoped, in fact, it could be used to prove convergence.

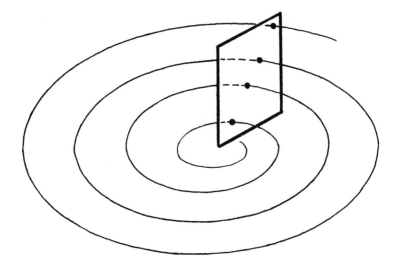

Poincaré section. A slice through the trajectories.

It didn't. Furthermore, he was unable to prove the series converged, or that the solar system was stable. Nevertheless he made impressive inroads, cracking the problem wide open with his 200 page paper. It impressed the judges so much there was unanimous agreement that he should be awarded the prize. Shortly after it was awarded, however, the French mathematician Edvard Phragmen noticed a mistake, and it appeared to be a critical one. Poincaré's paper had already been sent to press; it was going to be published in *Acta Mathematica*. The editor immediately stopped publication and seized all copies in print, then asked Poincaré to take another look at the problem to see if he could resolve the difficulty. Interestingly, it was during this second look that the breakthrough to chaos was made.

Poincaré went back to the problem with renewed vigor. He looked again at the patterns on the slices in phase space. Earlier he had only taken a cursory look at them. He concentrated on orbits that were close to periodic, but not exactly periodic. What

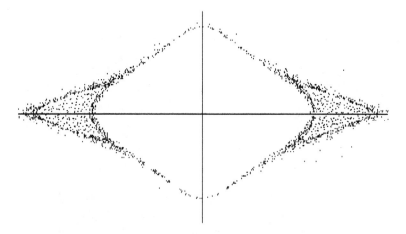

Poincaré section. Sections covered with large numbers of dots are regions of chaos.

did the pattern on the slice tell him? If the orbit was not periodic, patterns would not be repeated. But what would it do? With only a few passes it looked like the spray from a shotgun blast, but as the points continued to build up the patterns took on a strange appearance; some regions became dense with dots, other regions had none. It soon became clear to him that the orbit of a planet in a case such as this could not be calculated far into the future. In essence the series didn't converge—it diverged.

Poincaré was shocked by the results. He rewrote his paper outlining the new results. It was now 270 pages long, and was soon published in *Acta Mathematica*. Poincaré was so dismayed by the strange patterns that he didn't even try to draw them, they were just too complex. What he was seeing was the first glimpse of chaos.

With modern computers we can now much more easily duplicate what Poincaré only started to see. He looked at only a few dozen, or perhaps a hundred dots on his slice in phase space. We can easily look at millions, corresponding to millions of orbits of the planet. And what we see is an exceedingly strange pattern. Some regions are completely filled with dots, others contain none. There are islands, and islands within the islands. Particularly interesting are the patterns that are seen when things are magnified. The same strange patterns of islands, peninsulas, and so on are seen on a smaller scale. In fact if you magnify it further you see it again.

Poincaré didn't follow up on the problem, although there is no doubt that the strange result bothered him; he referred to it again and again throughout his life. But without a computer there was little more he could have done. There was also little interest from the scientific community. The main reason, no doubt, was that they weren't ready for the discovery. The methods Poincaré used were different and complex, and most mathematicians didn't know what to make of them. The three-body problem seemed to have encountered an impasse.

In 1913 Karl Sundham of Finland, however, did accomplish what the Oscar II contest asked for—a particular solution to the three-body problem expressed in convergent series. But the convergence was so slow the solution was next to useless.

Here was something new, something that had never been encountered in dynamical systems; it was unpredictable and chaotic—it was chaos. And soon it would be seen to pervade many other types of systems.

4

Building the Groundwork for Understanding Chaos

In the last chapter we had a brief look at chaos and at one of the most important tools at our disposal in studying it, namely phase space. Phase space plays such an important role in chaos it is essential that we sit down and take a closer look at it.

A CLOSER LOOK AT PHASE SPACE

Phase space is the space of dynamical systems, in other words, systems of one or more objects in motion. A good example of a simple dynamical system is a ball that is thrown into the air. What do we need to completely specify its motion? We need its height above the ground and its velocity (speed in a particular direction). Knowing these, we know everything there is to know about the system. They give the "state" of the system.

But we also want to know how this state changes in time, and we therefore need a law that can be expressed in mathematical form that predicts these changes. This law usually comes in the form of a differential equation. In the case of the ball, if we know its initial position and velocity, we can solve the differential equation and get an expression that will give its position and velocity at any time in the future or past.

This is the time-honored way of doing things, and for many years it was the accepted approach for all systems. Eventually, though, scientists and mathematicians discovered that you can't always get a simple analytical solution. This was the case that Poincaré and others encountered in the three-body problem. What did they do? They turned to geometry. I'm not referring to the geometry that you had in high school (Euclidean geometry), I'm using the word in a more general sense. They turned to pictures. Geometry gives us a way of turning numbers into pictures. This is basically what we are doing when we plot the path of the ball that was thrown. On the one hand, we can solve the equation and get a formula; on the other, we can plot the flight of the ball on a graph, and read off the coordinates from it. Either way we have a solution.

When scientists have a dynamical system that they can't solve analytically, they plot its motion in phase space. In plotting it they have solved the problem. In short, they have obtained what they want to know.

THE GUINEA PIG OF CHAOS

The usual starting point for studying chaos in the physical sciences is the pendulum. Chaos is, of course, not restricted to the physical sciences. It pervades all the sciences, and also other areas such as business and economics, but for now we'll restrict ourselves to the physical sciences.

The pendulum is such a simple device, it might seem that there is little to learn from it. After all, the only thing it can do is move back and forth. We'll see, however, that it's not as simple as it appears. There's a lot more to it than you may think.

Galileo was the first to recognize one of the pendulum's most important properties. He noticed that when you pull the bob to the side and let it swing, the time it takes to complete a swing is independent of how far you pull it to the side (its amplitude). If you pull it back a little further it goes a little faster, and completes the swing in the same time. Only if you change the length of the pendulum does the period, or time for one oscillation, change.

Actually, the above isn't exactly true. If you pull the pendulum bob back so it swings through a large amplitude, the time for a swing is not the same as that for a small amplitude swing. For small amplitudes, however, period is approximately independent of amplitude.

Galileo realized that the pendulum could be used to measure time. Throughout his life he was plagued with an inability to measure time accurately. Time was critical to many of his experiments, and the lack of an accurate clock frustrated him. Toward the end of his life he experimented with a pendulum clock, but was unable to perfect it.

The pendulum clock was soon perfected, however. The problem was that the amplitude died away; in other words, it got smaller and smaller until it finally stopped. A slight push at the end of each swing was therefore needed to keep it going, and the man who showed how this push could be applied was the Dutch physicist, Christian Huygens.

Born in The Hague in 1629, Huygens attended the University of Leiden. His early training was in mathematics, but he eventually became interested in astronomy and physics, and most of his contributions were made in these areas. He did, however, make one important contribution in mathematics; he wrote the first book on probability theory. His discoveries in astronomy were numerous: After helping perfect the telescope, he used it to discover Titan, the largest moon of Saturn, and he made extensive drawings of Mars. He is, in fact, credited with discovering the large dark region on Mars called Syrtis Major.

His major contribution, however, was his invention of the pendulum clock. At the time there were two devices for measuring the passage of time—both exceedingly crude. One was the water clock in which the flow of water was controlled and measured, and the other was a slowly falling weight. What was needed was a device for measuring short periods of time, and the pendulum seemed to be the key. Huygens showed that you could attach a weight on a pulley to the clock, and adjust it so that just enough energy was transferred at the end of the swing to keep the pendulum going. It was a major breakthrough, and pendulum clocks were soon used throughout Europe.

We will discuss several different types of pendulums in the next few sections, but before we start I should point out that the pendulum is really representative of a large class of objects, referred to as oscillators. Examples are a weight on the end of a spring, a ball rolling in a bowl, and a steel rod clamped at one end and pulled aside at the other. Each of these systems gives rise to phenomena similar to that seen in the pendulum.

Let's begin by distinguishing between real and idealized pendulums. In real pendulums, as we saw above, friction and air resistance damp the oscillations and eventually stop it. In the idealized pendulum there is no friction or air resistance and the bob swings at the same amplitude forever. No such object exists in nature, but this doesn't stop us from considering it.

We can, of course, get around damping as Huygens did, by giving the bob a little push at the end of each swing, but this is actually a different kind of pendulum, one called the forced pendulum.

We'll begin with the simple idealized pendulum.

THE PENDULUM IN PHASE SPACE

One of the reasons we deal with the pendulum is that it is easy to plot its motion in phase space. If the amplitude is small, it's a two-dimensional problem, so all we need to specify it completely is its position and its velocity. We can make a two-dimensional plot with one axis (the horizontal), position, and the other (the vertical), velocity.

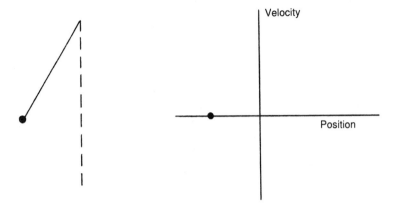

Pendulum at the beginning of its swing.

Let's assume we pull the bob to one side and take a picture of it every 1/10th second. As we hold the bob to the side it is a finite distance from its equilibrium position, but it has no velocity. The point representing this state is to the left in the phase diagram. When we let the bob go its position changes—it decreases—and its velocity increases from zero to a finite value. How do we determine its new velocity and position? They come from the differential equation that describes its motion.

As the bob moves we get a new point, or new state in phase space, as shown in the diagram on the following page. Incidentally, phase space is sometimes referred to as state space because it is a plot of all the states of the system. As the bob continues to move it passes through a continuum of states, until finally at its equilibrium position it has maximum velocity.

Now, let's see what happens after it reaches the bottom of the swing. Its position, which until now has been on the negative side of the axis, becomes positive, and its velocity begins to decrease. Finally it stops at the end of the swing, and it starts back again. In short, it completes a loop in phase space and does this

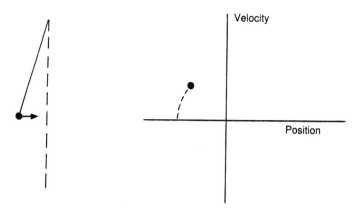

Pendulum just after beginning its swing. Its velocity is low. The two numbers (position and velocity) specify a point in phase space.

over and over again. We refer to this track as a trajectory; if it is closed as it is in the above case we refer to it as an orbit.

Once we have plotted the trajectory, the state of our system can be determined at any time by reading the coordinates. It might seem that this is not much of an advantage; after all, we can write

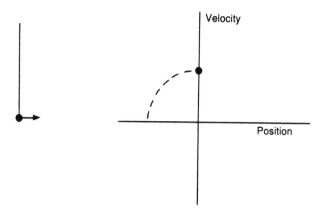

Pendulum at the bottom of its swing. Its velocity is maximum.

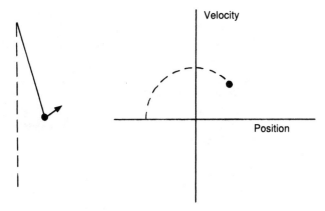

Pendulum past bottom of swing. Its velocity is decreasing.

down a formula for the simple pendulum and have all the information we need. As I mentioned earlier, however, you can't always solve the problem analytically and get a simple formula.

Now, let's pull the pendulum bob back a little further and let it swing from side to side again, and make a plot as we did

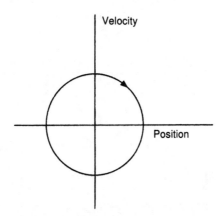

Complete trajectory of pendulum in phase space.

earlier. We get a circle again, but this time it is larger. We're pulling it further to the side so the initial position is greater, and as it moves its velocity becomes faster and is greater as it passes through the equilibrium position.

If we continue doing this for larger and larger amplitudes, we get a series of circles, one inside the other. We refer to this as a phase portrait. When we have a trajectory in phase space we refer to it as a phase diagram. A group of trajectories, on the other hand, is usually called a phase portrait.

Note that there is a direction to the lines we have plotted in phase space, so it is convenient to put arrows on them. We can therefore visualize a sort of "flow." As I emphasized earlier, however, this is an idealized case; it doesn't occur in nature. In nature there is friction and it has an effect on any system we may be dealing with. We refer to such a system as dissipative. We now ask: What does the trajectory of a real system look like in phase space?

Again, we'll pull the bob to the side and let it go. As we watch, we'll see that while it oscillates the amplitude gradually decreases until finally the bob settles at its equilibrium position. In short, distance and velocity get smaller and smaller. If we plot this in phase space we get a spiral that ends at the center, something quite different from the idealized case.

All of this may seem rather tedious. When are we going to get to something exciting, you ask? What has this got to do with chaos? At this stage all I can say is: be patient. We will see that it will be of tremendous help in understanding some of the most fascinating aspects of chaos.

NONLINEARITY

Even though I haven't emphasized it strongly, everything we've done so far is approximate. I mentioned earlier that Galileo showed that period was independent of amplitude, but this is only valid for small amplitudes, and even then it's an

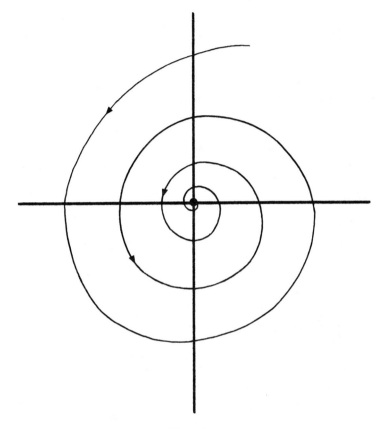

A sink in phase space.

approximation. With this approximation the pendulum is a simple problem, but if you try to solve it without making the approximation you'll soon find that it's actually a very difficult problem. The pendulum is a nonlinear system. What we are doing in making the approximation is assuming that it is a linear system.

This applies to all oscillating systems. When you make a plot of linear motion you get a straight line. For small amplitudes most oscillators are linear; as the amplitude increases, however, the line begins to curve and the oscillator becomes nonlinear. Linear systems are described by linear equations, nonlinear systems by nonlinear equations. Also, when an equation is linear we can take two solutions and add them together and get another valid solution. Linear equations are therefore usually relatively easy to solve, or at least they're solvable. Nonlinear equations, in general, are not; in other words, we can't get a simple analytic solution. As anyone who has taken an undergraduate physics course knows, most of mathematical physics centers around linear equations. For years nonlinear equations were ignored; they were either too hard to solve, or didn't have a solution, so the attitude was: Why bother with them? For the first half of this century (and even earlier) a physicist's training centered around linear equations, and by the time they graduated they were quite proficient at solving them. Nature was basically linear, they were told, and anything that wasn't could be ignored. Gradually, though, in the 1970s and 1980s scientists began to realize they were wrong. Everything in nature wasn't linear; in fact most systems were nonlinear. Linear systems were the exception.

This was a shock, and required a new way of looking at the world. Furthermore, it applied not only to nature but also to other areas such as economics, social sciences, and business. The real shock, however, wasn't that scientists were now up against systems they couldn't find analytical solutions for. After all they could still resort to phase space, but nonlinear equations gave rise to something that never occurs in linear systems, namely chaos.

Chaos wasn't new; it had been observed for years, but nobody had paid much attention to it. All you have to do to see chaos is go to the nearest stream, assuming there are a few rocks in it, and see that it is flowing relatively fast. The turbulence you see in the water is chaos.

When scientists were working primarily with linear equations, they didn't have to worry about chaos. But when they realized that most systems in nature (e.g. the weather, fluid motion) were nonlinear and chaos arose in nonlinear systems, they knew they would have to take a different approach.

To see what this approach was, let's return to the pendulum. The real pendulum, as distinct from the idealized one, is nonlinear. Again, let's plot its trajectory in phase space, but this time we won't worry about keeping the amplitude small; in fact, we'll go to very large amplitudes.

If we do as we did before, we get ovals (or ellipses) in phase space around the center point, instead of circles. As we go to greater and greater amplitudes we get larger and larger ovals. This time let's go a lot further than we did previously. In fact, let's go as far as starting the bob right at the top (we would, of course, have to assume the string was rigid). We find when we let it go it swings right around and comes to the top again on the other side. Our diagram in phase space looks as follows:

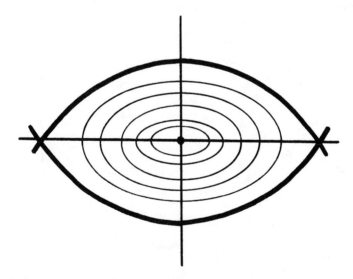

Phases portrait for a nonlinear pendulum. All the trajectories for different amplitudes in phase space.

Can we go any further? Indeed, we can. You don't normally think of a pendulum as something that goes round and round, but you could have the bob pass over the top and keep going. It would be like a boy with a slingshot, whirling it around with a rock in the pouch. If we include trajectories of this type in our diagram we see they are not closed, so they are not true orbits. In fact, you can have two types of trajectories of this type: one going clockwise and the other going counterclockwise. Adding them into our diagram we get the following:

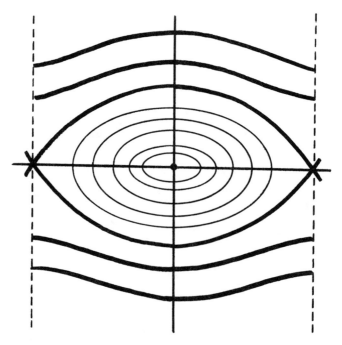

An extension of the previous diagram to include trajectories that go all the way around.

Notice in this diagram there are two trajectories that cross; they form cusps near the ends of the picture. They are referred to as separatrices and are important because they separate the two types of motions of the system. Inside the separatrices we have back and forth oscillatory motion. Outside there is continuous round and round motion in either the clockwise or counterclockwise direction.

I may have given the impression that the above case is representative of the "real" pendulum because we are treating it nonlinearly, and real problems are nonlinear. But we are still dealing with the idealistic case in that there is no damping. Real pendulums are damped as a result of friction. To get the trajectories in phase space that we got in this diagram we would have to give the pendulum a little push at the end of each swing as Huygens did. This is referred to as the forced pendulum. The phase portrait for a forced pendulum would look similar to that shown above.

WRAPPING UP PHASE SPACE

Let's return to the above phase diagram. As I mentioned earlier the separatrices separate the two types of motion. Consider the points along the separatrices where they cross; there is one to the right and one to the left. Physically they correspond to the pendulum standing straight up. The one on the left corresponds to the pendulum moving to the left until it is finally directly upward; similarly the one on the right corresponds to motion to the upward position on the right.

The important point here is that for both cases the pendulum is in the same position (the bob is at the same point). Yet on the diagram these points appear to be widely separated. How can we patch up this apparent defect? The easiest way is to wrap the phase diagram around a cylinder so that the right side matches the left. This tells us immediately that they are the same points. Notice that we can't do this in the other direction, namely

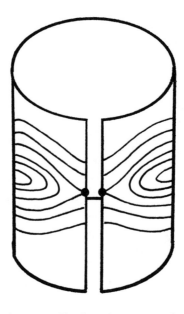

Wrapping it up in phase space. The above phase portrait plotted on a cylinder.

the velocity direction, because we are dealing with velocities in opposite directions.

HIGHER DIMENSIONS

In selecting the pendulum to illustrate the ideas of phase space, we have selected a particularly simple system. Only two dimensions are needed to specify a state, so we can easily draw its phase diagram on a piece of paper. The real power of phase space, however, comes when we apply it to more complex systems, and it doesn't take much more to make the system complex. As we saw earlier, even a system of three bodies is complex.

What do we do if we have several objects in our system? For a complete solution we need the state of each object, and this means specifying its position and velocity. If we had three objects and each of them was in three-dimensional space, we would need a space with 18 dimensions.

Higher dimensional spaces, however, are no problem. Mathematicians and scientists have been using them for years. Einstein, for example, used a four-dimensional space (more exactly, spacetime) in his theory of relativity. It's difficult, if not impossible, to visualize more than three dimensions, but we can try. Let's see how we would go about it. Starting with a point, which has zero dimensions, we can create a one-dimensional space by moving it sideways. Our one-dimensional space is a line. If we now move the line perpendicular to itself we get a two- dimensional space. And finally if we move the sheet perpendicular to itself we get our usual three-dimensional space.

How would we go further and get a four-dimensional space? Obviously we would have to move the three-dimensional space perpendicular to itself. It's hard to see exactly how you would do this. You could expand it sideways, but it would probably be more meaningful to expand it outward as shown in the figure.

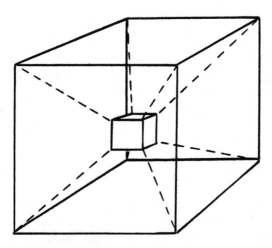

A simple representation of four dimensions.

Again if we wanted a representation of a five-dimensional space we would have to move this perpendicular to itself. It's easy to see from this that higher dimensional spaces are difficult to visualize. Fortunately, we don't need to visualize them. We can easily set them up mathematically and work with the resulting mathematical expressions. Scientists have been doing it for years. This means we can easily deal with three or even a large number of objects in phase space; it will be a higher dimensional space, but that's not a problem.

SINKS AND SOURCES

For complex systems, phase portraits can be complex. We have already seen several of the features that can occur. Let's go back to the damped pendulum. We got a phase diagram that looked as shown on the following page.

In short, the trajectory spiraled into a point. If you visualize the flow as water, it reminds you of the whirlpool pattern you get when water goes down a sink. We therefore refer to it as a sink. The trajectories in a sink don't have to spiral in as they do in the above case; they can come in directly, and there can be several of them.

It's important to note here that sinks are stable. This means that if the point representing the system is pulled a short distance away it will come back to its initial position.

It is, of course, also possible that the flow could go in the opposite direction. In this case you would have an outward flow—a source. Again, the lines need not spiral out, and there can be several trajectories coming out. In contrast to the sink, states here are unstable in that if you displace them slightly they move away.

Looking again at the phase diagram for the nonlinear pendulum you see another phenomenon—lines crossing, or at least appearing to cross. As we saw earlier this corresponds to the

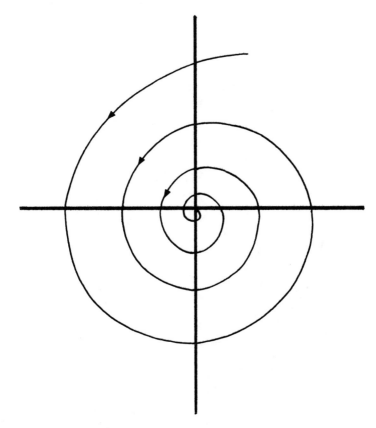

A sink in phase space.

pendulum sitting straight up. We refer to these as saddles, and the center point as a saddle point.

In our diagram they occur along the separatrices. In this case, states along the incoming trajectories are stable; states along the outgoing ones are unstable.

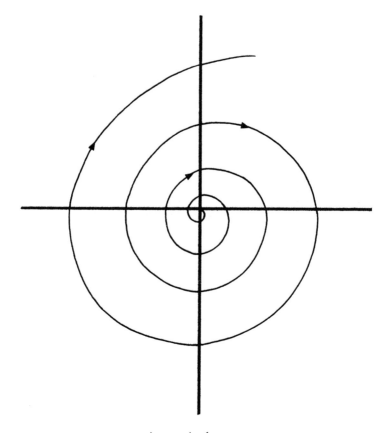

A source in phase space.

ATTRACTORS

In the case of the sink we saw that all the trajectories lead to a point. What this means in relation to the pendulum is that the amplitude of the oscillations gets smaller and smaller, and eventually the bob stops. One way of looking at this is to say

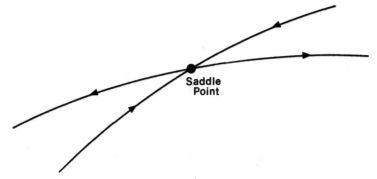

A saddle point in phase space.

that the pendulum is attracted to the point where it stopped. We refer to it as a fixed-point attractor.

Fixed-point attractors can easily be spotted in a phase diagram; they are associated with sinks. In any given diagram there may be one or more fixed point attractors, or there may be none.

What about other types of attractors? To look into this, let's go back to the forced pendulum, or more generally, the forced oscillator. A good example, incidentally, is your heart. The Dutch physicist, Balthasar van der Pol was examining a mathematical model of the heart in the 1920s when he came upon another attractor. He later found the same attractor was associated with electronic vacuum tubes.

The attractor he discovered is called the limit cycle and we can use a forced pendulum, or more explicitly, a grandfather clock, to illustrate it. The phase space trajectory for the pendulum of a grandfather clock is an orbit. It isn't necessarily a circle, but it will be closed. If you pull the pendulum a little too far it will still settle into its regular period. In phase space we represent this as shown on the following page.

The ellipse is the usual orbit of the forced pendulum. If the initial state is just outside it, as shown, it is attracted to this ellipse. Similarly, if it is just inside, it is also attracted. So we have

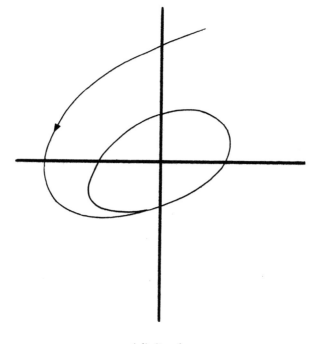

A limit cycle.

another attractor—the limit cycle—an attractor that is associated with periodic motion.

Are there other types of attractors? Indeed, there are. If we superimpose two attractors of the above type, in other words, two limit cycles, we get another type. What does it look like? A moment's reflection will tell you that it is a torus; in other words, a surface like that of a tire. One period is associated with the short circumference around the tire, and the other with the long circumference which is perpendicular to it.

Of particular interest in relation to this torus attractor is quasiperiodic motion. Since there are two limit cycles in this case, there are two periods. Furthermore, if the ratio of the two

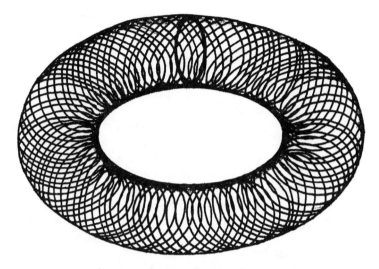

A torus. Quasiperiodic motion takes place on a torus.

periods is an integer the combination is periodic, but if the ratio is not an integer you get quasiperiodic motion. This is motion that never exactly repeats itself, nevertheless is regular and predictable. The trajectory will go around both the small radius of the torus and the large one. It will wind around and around without ever coming back to the same point.

None of the attractors we've talked about so far give rise to chaos, but we'll see in the next chapter there is another type that does.

A system may have one or more attractors in phase space. You can have several of the same type, or several of different types. An example of a system with two types is a grandfather clock. If you pull the pendulum a very short distance from its equilibrium position, its oscillations quickly damp and it will stop. It stops at a fixed-point attractor. If you pull it aside a little further, however, it will start clicking off seconds and work as a clock. In short, it will be attracted to its limit cycle. We refer to the set of points around each of these attractors as the basin of attraction.

BIFURCATION

We will have a lot to say about bifurcation later, but this is a good place to introduce it. This time we won't plot the motion in phase space, we will begin by considering the stability of the motion. The stability depends on certain parameters in the system (e.g., period) so we make a plot using the parameters as axes. Over a certain range of the parameters the system is stable; over another range it is unstable. Note that the marble in a) is in unstable equilibrium; a slight shove and it goes a long distance. The marble in b) is in stable equilibrium.

One form of bifurcation occurs when you go from a stable region to an unstable one; a significant change occurs here. The point at which this occurs is called a bifurcation point.

Generally, any significant change at a fixed point in a graph is called a bifurcation. The period, for, example, may change at

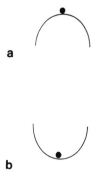

a

b

Marble in the upper diagram (a) is in an unstable situation. One in the lower diagram (b) is in a stable situation.

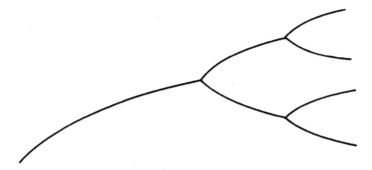

Branches emerging from bifurcation points.

the bifurcation point. Several branches may emerge from this, each representing a solution of the equation.

With this we are now in a position to introduce one of the basic things that characterizes chaos, namely the strange attractor.

5

Strange Attractors

Three types of attractors were introduced in the last chapter: fixed-point, limit cycle, and torus. Each is associated with a different type of motion, and each is important in dynamics. The understanding of attractors was an important milestone. In this chapter we will see that there is another type of attractor, called the strange attractor, and it will play a key role in chaos theory.

Before we discuss strange attractors, however, let's go back to dynamical systems. Scientists have shown that they are of two types: dissipative and conservative. In a dissipative system energy is lost, usually via friction, as in the case of the pendulum discussed earlier. In a conservative system, on the other hand, there is no energy loss. The planets of the solar system are an example of a conservative system. Their orbits remain virtually the same over hundreds of thousands and even millions of years. Particles in electromagnetic fields and plasmas are other examples of conservative systems.

One of the signatures of a dissipative system is contraction. For example, as a pendulum loses energy, the amplitude of its swing decreases. The same thing happens in astronomical systems, but over a much longer period of time. It is so small in most cases that it is usually neglected and the system is treated as conservative. But if you look closely at a binary star system, for example, you see that the two stars slowly lose energy and orbit closer and closer to one another. If you are interested in the long-term behavior of such a system you have to treat it as dissipative. This also applies to galaxies and clusters. Over a long period of time stars are lost due to collisions and the system shrinks. Dissipative effects are also important in the formation of the solar system, in the formation of the universe, and in cosmology in general.

Chaos appears in both dissipative and conservative systems, but there is a difference in its structure in the two types of systems. Conservative systems have no attractors. Initial conditions can give rise to periodic, quasiperiodic, or chaotic motion, but the chaotic motion, unlike that associated with dissipative systems, is not self-similar. In other words, if you magnify it, it does not give smaller copies of itself. A system that does exhibit self-similarity is called fractal. A good example of a fractal, in case you're not familiar with them, is an ocean coastline. If you look at it from an airplane, it is jagged, with many inlets and peninsulas. If you look at a small section of it closer, you see the same structure. In other words, it's still jagged. In a fractal you can, in fact, continue to magnify it indefinitely and still see the same structure.

The chaotic orbits in conservative systems are not fractal; they visit all regions of certain small sections of the phase space, and completely avoid other regions. If you magnify a region of the space, it is not self-similar.

A major breakthrough in understanding chaos in dissipative systems was the discovery of the strange attractor, and the story of this discovery begins with Edward Lorenz.

LORENZ AND THE WEATHER

Edward Lorenz loved weather. Even as a youth in West Hartford, Connecticut, he kept a diary of changes in the weather,

Edward Lorenz.

and maximum and minimums in temperature. But he was also fascinated by puzzles, or more generally, mathematics. He loved to work out mathematical puzzles. They were challenges—challenges he couldn't pass up—and he spent a lot of time with them. He eventually became so proficient at solving puzzles he decided to become a mathematician. Before he could realize his dream, though, World War II came and he was inducted. There was little need for mathematicians in the branch of the service he selected, the Army Air Corp, but there was a need for meteorologists, and before long Lorenz was working on the weather. Weather, he soon discovered, was just as much a puzzle as many of the mathematical puzzles he had worked on, and he soon became intrigued with it. When he was discharged he therefore decided to stay with weather. He went to Dartmouth College, where he received a degree in meteorology.

Deep down, though, he was still a mathematician, and a few years later when he was at MIT he saw an opportunity to couple mathematics and meteorology. Weather prediction was a problem; meteorologists could predict the weather for a few days, but beyond that it was hopeless. Was there a reason for this? Why was the weather so unpredictable? The way to find out was to make a mathematical model of the weather; equations that represented changes in temperature, pressure, wind velocity, and so on. Weather was so complex, however, that it was difficult to model. Lorenz finally found a set of 12 equations that contained such things as pressure and temperature, and with them he was able to set up a crude model. The calculations needed, even with this crude model, however, were extensive, but luckily computers were just coming on the market. In 1960 they were still monstrosities—made up of hundreds of vacuum tubes that easily overheated—and breakdowns were common. But when they worked they were miraculous, performing hundreds of calculations per minute.

Lorenz's computer was a Royal McBee—slow and crude by today's standards, but indispensable to Lorenz. He watched eagerly hour after hour as numbers poured out of it representing various aspects of the weather. It was an intriguing machine, generating weather day after day, weather that changed, weather that never seemed to repeat itself. But eventually Lorenz became dissatisfied; he wasn't learning as much about long-range forecasting as he had hoped, so he simplified his set of equations. He concentrated on convection and convective currents—one aspect of the weather. Convective currents are all around us. Hot air rises; cool air descends; and it happens in the atmosphere every day, giving rain, snow, wind, and other things. Lorenz's convection currents were circular; cold air from the top of the atmosphere was descending in one point of the circle; hot air from near the Earth was rising in the other part.

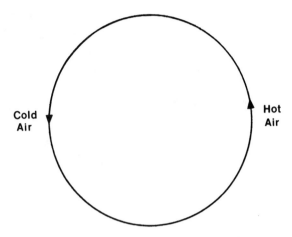

Convection cycle showing hot air rising and cold air descending.

He arrived at three seemingly simple equations representing convection. Any mathematician looking at them would say, "I can easily solve those." Simultaneous equations in three variables are, indeed, solved by students in high school. But these weren't algebraic equations; they were differential equations, and despite their simple appearance they were complex.

Lorenz set them up on his computer, and again numbers came pouring out that represented some aspect of the weather. He plotted the points and got a continuous line. Lorenz's aim at this point was to find out if long-range forecasting was possible. Large memory, fast computers were just starting to be built, and satellites had been put in orbit a few years earlier. It appeared to be the dawn of a new age. World-wide weather forecasting was a possibility, and with large enough computers, forecasts for months ahead could be projected. It was the dream of many. Was there anything standing in the way? Lorenz was sure his model of convection in the atmosphere would answer this question and perhaps help clarify any problems that might

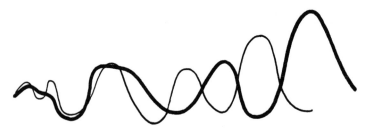

Sensitive dependence on initial conditions. Two systems (represented by a heavy line and a light one) start out with slightly different initial conditions. Within a short time their trajectories are entirely different.

come up, but he had no idea at the time how important it would eventually prove to be.

Day after day he examined the patterns that came out of his computer; they appeared to be random. Looking at a pattern one day in 1961, he decided to repeat it. Taking the numbers that came out, he put them back in as initial conditions, expecting that the run would repeat itself. And, indeed, for the first while (a few days in real time) it was close, but it soon began to diverge, and to his dismay the two lines were soon so far apart there was no resemblance between them. Not only was the original output not duplicated; it wasn't even close. He tried again with the same result. Maybe there was a slight error in the initial data, he told himself. In other words, the ones he had put in may have been slightly different from the original ones. But even if this was the case, it had been accepted since the time of Newton that small errors cause only small effects. In this case the effect was huge; in fact, after a short time there was no resemblance whatsoever to the original output.

Lorenz wondered how an error could occur, then he realized that the calculations within the computer were performed using six significant figures, but it printed out only three significant figures (e.g., .785432 and printed out .785). He had taken the

The Lorenz attractor (butterfly diagram). Projection on the zx axis. (Matthew Collier)

three digit numbers and put them back into the computer, introducing an error of a few thousandths.

But how could such a small error cause such a dramatic effect? Lorenz knew he was on to something important. The overall figure that he got in phase space was also a surprise; it looked like the wings of a butterfly. After plotting several thousand points he got two loops that resembled the left wing and five loops that resembled the right wing. This is now called the butterfly effect. It was easy to see that a point in phase space moving around the loops would never repeat its motion. It might go around the left loop, then twice around the right, before going back to the left. Where it went around the loop and which loop it chose to go around could not be predicted. Its motion was random, or chaotic.

Lorenz published his discovery in the *Journal of Atmospheric Science* under the title "Deterministic Nonperiodic Flow." But for almost a decade nobody paid any attention to it. To Lorenz, however, it was an important breakthrough in understanding long-range forecasting of the weather. If tiny changes in the initial data introduced significant changes in the weather only a few days away—changes that appeared to be random—he knew that long-range forecasting was doomed. It would be impossible to forecast the weather a month in advance.

Meteorologists found Lorenz's article interesting, but the mathematics in it deterred them from looking further, so nobody followed up on it. Mathematicians and physicists would no doubt have been interested, but they didn't read the *Journal of Atmospheric Science.*

What Lorenz had discovered was the first strange attractor. It wasn't called a strange attractor until many years later, but it was the first of many that would eventually be discovered.

Weather, as it turns out, is only one of several places where chaos arises. As I mentioned earlier you can see chaos any day of the year by merely turning on your water tap. When the water runs slowly, it is smooth and clear. Turn it on fast and you see a significant change: It becomes turbulent. Is turbulent motion chaotic? Indeed, it is, and as we will see a theory of turbulence had already been proposed many years before Lorenz began his work.

LANDAU'S THEORY OF TURBULENCE

Long before the rest of the world began to take an interest in turbulence, the Russians had turned their attention to it.

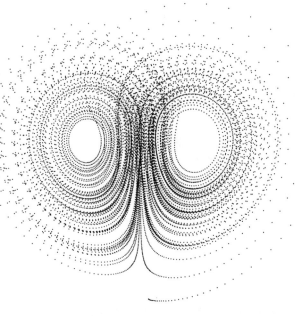

Lorenz attractor. Projection on the yz axis. (Matthew Collier)

Andrei Kolmogorov had put forward a theory of turbulence as early as the 1930s. He suggested that turbulence was associated with eddy currents. According to his theory, eddies formed within eddies, cascading rapidly to smaller and smaller eddies.

Lev Landau, also of the USSR, studied Kolmogorov's theory and found it incomplete. It said nothing about the onset of turbulence, and to Landau this was a particularly important feature of the phenomenon.

Landau was born in Baku in 1908; his father was an engineer and his mother a physician. He studied at the University of Baku, then later at the University of Leningrad where he graduated in 1927. He then travelled throughout Europe studying and in 1934 received his Ph.D. from the University of Kharkov.

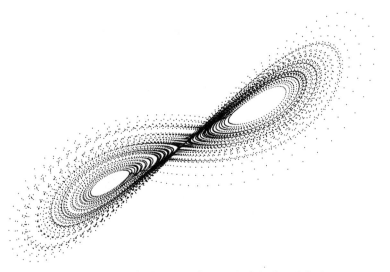

Lorenz attractor. Projection on the xy axis. (Matthew Collier)

Much of Landau's research was on magnetism and low-temperature physics. He was awarded the Nobel Prize in 1962 for his work in low-temperature physics. Along with E.M. Lifshitz, he wrote a series of books on graduate-level physics familiar to all physics students; they covered literally all areas of the discipline. One of the books was on fluid mechanics.

In January, 1962, Landau almost lost his life in an auto accident near Moscow; it left him with eleven broken bones and a fractured skull. He hovered between life and death for months, and spent almost two years in the hospital. He never fully recovered from the accident, and died a few years later.

Landau's interest in turbulence began in the early 1940s, and by 1944 he had published one of the classic theories in the area—a theory for the onset of turbulence. A few years earlier a fellow Russian, Eberhard Hopf, had put forward a theory for the creation of "wobbles" in fluid flow, a process that is now known as Hopf bifurcation. Landau's theory was an extension of Hopf's theory.

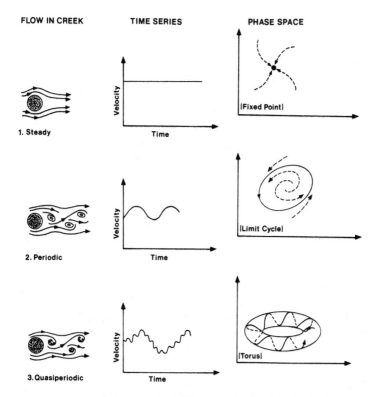

The transition to turbulence. A simplified version showing behavior in phase space.

The easiest way to understand Landau's theory is to imagine a rock in a stream where the flow rate can be adjusted. Assume that we put a velocity gauge slightly downstream from the rock. When the flow is slow there is a steady state and the gauge registers a constant velocity. This corresponds to a fixed-point attractor. Now let's assume we speed the water up slightly and an eddy forms behind the rock, an eddy that eventually moves downstream. As the eddy passes the velocity gauge it will register an increase in the rate of flow. Then another eddy forms and does the same thing. The motion of the water becomes periodic, and the attractor has become a limit cycle. A bifurcation has occurred as the water has moved from steady state to periodic flow. Let's assume the periodic flow has frequency f_1.

Now, assume we increase the speed of the stream a little more. Smaller eddies will now appear with a different frequency f_2; they will be superimposed on f_1, and the system will have gone through a second bifurcation. Since there are now two different periods, the attractor will be a torus. As we saw earlier the motion on this torus can be either periodic or quasiperiodic, depending on the ratio of the frequencies. If they are integral the mode will be periodic; if not, it will be quasiperiodic.

Let's turn now to Landau's theory. According to it, as the flow rate of the fluid increases, quasiperiodic modes form, and eventually motion on the torus, which at this stage is two-dimensional, becomes unstable. Small disturbances as the water velocity increases lead to new eddies of a different frequency, and this in turn leads to quasiperiodic motion on a three-dimensional torus. This trend continues as eddies of different frequency form within the existing eddies. We get four-dimensional quasiperiodic motion, then five-dimensional, and so on. In theory this continues indefinitely until there is an infinite number of modes of different frequencies present. Real turbulence appears to be made up of large numbers of different frequencies, and because of this Landau's theory was accepted for over 30 years. It wasn't challenged until the late 1960s.

A SIMPLER THEORY

David Ruelle of the Institut des Hautes Etude Scientifiques near Paris studied Landau's theory and soon became dissatisfied with it. It was too complex.

Born in Ghent, France in 1935, Ruelle was the son of a gym teacher and a professor of linguistics. Like many mathematicians he enjoyed hiking and backpacking, and frequently went on extended backpacking trips. It gave him time to think, he said. Unlike most mathematicians, however, he was interested, not only in pure mathematics, but also applied mathematics.

In 1968 he became interested in fluid flow and began teaching himself hydrodynamics by reading Landau and Lifshitz's book *Fluid Mechanics*. "I worked my way slowly through the complicated calculations that those authors seemed to relish, and suddenly fell on something interesting: a section on the onset of turbulence, without complicated calculations," he wrote in his book *Chance and Chaos*. Upon reading the section he found it disturbing. "The reason I did not like Landau's description of turbulence in terms of modes is that I had heard seminars by Rene Thom and had read a paper by Stephen Smale called 'Differentiable Dynamic Systems.' " Thom was a colleague of his at the Institute where he worked, and Smale had visited it several times. Smale, in particular, had pointed out that there should be a simpler dependence on initial conditions in turbulence, and there was no such thing in Landau's theory. Furthermore, Ruelle couldn't see how this simpler dependence could be incorporated.

"The more I thought about the problem, the less I believed Landau's picture," Ruelle wrote. He was sure that if there were an infinite number of frequency modes in a viscous fluid, as proposed by Landau, they would interact more strongly and produce something quite different from turbulence.

Ruelle teamed up with a Dutch mathematician, Floris Takens, and together they attacked the problem. Ruelle and Takens showed that only three independent motions, not an infinite number as Landau postulated, were needed to create turbulence. But they needed something else, something they called a "strange attractor." (Unknown to them at the time, Lorenz had already discovered one several years earlier.) With this they were able to show how turbulence would begin.

The two men wrote a paper titled "On the Problem of Turbulence" and sent it to a respected journal in the field. Several weeks later they got a reply: The paper was rejected. The referee didn't like their idea, and was sure they didn't understand turbulence. He sent them some of his papers on turbulence to straighten out their misunderstanding.

At the time, Ruelle was an editor of a European journal, so he submitted the paper to himself, read it, and accepted it—something he said he did very cautiously. It was soon printed and is now considered to be a classic paper in chaos.

Years later when Ruelle learned about Lorenz's attractor he was delighted.

DETAILS OF THE STRANGE ATTRACTOR

With the strange attractor there were now four types of attractors, but the only one that gives chaos is the last. What is a strange attractor? From a simple point of view we can say it is an endless path in phase space where the future depends sensitively on the initial conditions.

More formally we can define a strange attractor as something that has the following four characteristics:

1) It is generated by a simple set of differential equations.
2) It is an attractor and therefore all nearby trajectories in phase space converge toward it.
3) It has a strong or very sensitive dependence on initial conditions. In short, tiny differences or errors in the initial conditions lead quickly to large differences in the trajectory.
4) It is fractal.

The name "strange attractor" was given to the objects by Ruelle and Takens. (They still have a friendly argument going on about who actually came up with the name first.) The word "strange" comes from the fact that they are based on contradictory effects. First, they are attractors, and therefore trajectories must converge to them. But at the same time they exhibit sensitive dependence on initial conditions, which implies that trajectories that start out together on the attractor have to diverge rapidly. It may seem that these two things can't occur at the same time, but as we will see, in the strange attractor they do.

Before we look into the resolution of this problem, however, let's consider the dimension of a strange attractor. If you look at the Lorenz attractor it appears as if several of the trajectories cross. But trajectories in an attractor can't cross; this would lead to a contradiction. The system of equations is deterministic and if the trajectories crossed it would mean that the system would have a choice at the intersection; it could go one route or the other, and this can't happen. This means the attractor has to consist of a stack of two-dimensional sheets, so that one path is actually passing behind the other. This, in turn, implies that the dimension of the attractor must be greater than two. Also, in the case of the Lorenz attractor, as with all dissipative (energy losing) systems, any specified area of initial conditions contracts in time, which implies the dimension of the attractor must be less than three. The dimension of the Lorenz attractor is therefore between two and three; in other words, it is not an integer. Systems with noninteger dimensions are fractals, which, we have seen, is one of the properties of the strange attractor.

THE HÉNON ATTRACTOR

About the time Lorenz was working on the weather, an astronomer in France, Michel Hénon, was working on an astronomical problem related to star clusters. A strange attractor also came out of Hénon's work and, as in the case of Lorenz's, it was years before the significance of Hénon's attractor was realized.

Michel Hénon was born in Paris in 1931 and became interested in science at an early age. He was particularly interested in astronomy, but also intrigued by mathematics, so when he selected a thesis on astronomy it was a mathematical problem—a problem involving the dynamics of stellar motions. To Hénon one of the most intriguing objects in astronomy were globular clusters; they are clusters containing from a few hundred thousand stars to a few million. Although they looked like miniature galaxies, they are, in reality, quite different. Composed entirely of old red stars, they are put together haphazardly in that the stars within them orbited randomly around the center of the object. In one respect they are like our solar system. In our solar system the planets orbit the sun, and the gravitational force on them is directed toward the sun because most of the mass of

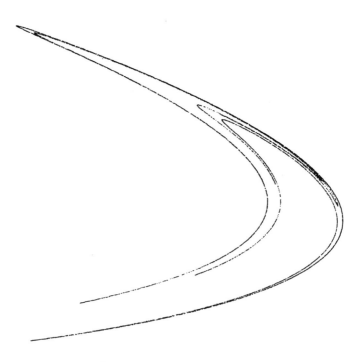

The Hénon attractor. (Matthew Collier)

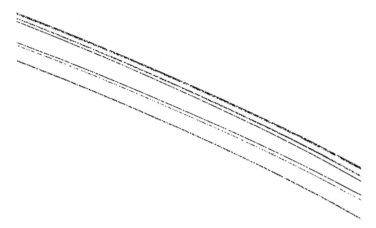

Section of the Hénon attractor. Magnified ten times. (Matthew Collier)

the solar system is in the sun. Gravity is also the dominant force in globular clusters; however, the gravitational source is not a point, but a disk with thickness in three dimensions.

Hénon became interested in the orbits of stars in globular clusters and decided to do his doctoral thesis on them in 1960. With hundreds of thousands of stars in a globular cluster it was obviously an extremely difficult problem, and a large number of simplifications were needed. One of the first things he found was that the core of the cluster would collapse over a very long period of time. Stars would collide and fly off into space, and as a result the cluster would lose energy and collapse. Over billions of years the cluster would shrink, but it would remain similar. Such a system is dissipative. As I mentioned earlier, most astronomical systems are not assumed to be dissipative, but over a long period of time they do lose energy. Hénon's calculations showed that this collapse would continue indefinitely, and the system would seek a state of infinite density.

Hénon left the problem for a while, but in 1962, while he was at Princeton University, he came back to it. He now had access to computers and could do more. Along with a Princeton graduate student, Carl Heiles, he set up a system of equations for the orbits of the stars in the system, and reduced them to the simplest possible form. After calculating the orbits in phase space, Hénon did something similar to what Poincaré did years earlier: He looked at the stars as they passed through a two-dimensional sheet. The orbits were strange; some appeared to be distorted ovals, others were figure eights. What was particularly strange, though, was that the curves were not closed; they didn't come back to exactly the same place, and they never repeated themselves.

Also amazing was that some of the orbits were so unstable that the points scattered randomly across the page. In some places there were curves and in others there were only random points. In short, there was order mixed in with disorder.

Hénon finished the work, published it, and went on to other problems. Eventually he moved to Nice Observatory in the South of France. In 1976, he heard about the strange attractors of Lorenz and Ruelle, and began wondering if they were related to the work he had done on globular clusters.

He went back to the problem again. This time he worried less about the astronomical aspects of the problem and concentrated more on the equations. He simplified them as much as possible. They were now difference equations rather than differential equations, and they were simple, even simpler than Lorenz's equations. He also had access to computers so he could plot millions of points. What he got was something that resembled a banana. As the program ran the shape became increasingly clear, and more detail appeared. He magnified it and looked closely at it; it had a substructure—similar to the original structure. As he magnified it more and more he noticed it was self-similar. It was fractal.

What Hénon had discovered was another attractor. Scientists hoped at first that it was simple enough to give significant insight into the nature of strange attractors, but it was still complex enough that little was learned. It had a fractal dimension

Stephen Smale. (G. Paul Bishop, Jr., Berkeley, California 94704)

between one and two, whereas Lorenz's attractor had a dimension between two and three.

Hénon realized that something odd was going on in phase space to give such an attractor. The space was being stretched and folded. Interestingly, such a mapping had already been suggested by Stephen Smale.

STRETCHING AND FOLDING PHASE SPACE

In the early 1960s, Stephen Smale of the University of California at Berkeley became interested in dynamic systems. A pure mathematician, he approached applied problems differently from most people. By 1960 he had already won honors for his contributions to topology—a branch of mathematics that is concerned with shapes in space. What would happen if space were rubber and could be twisted and distorted? This is the type of question Smale would ask himself when examining the topological properties of a system.

Smale had always worked on pure mathematical problems, but he wanted to expand his horizons. What could he apply his knowledge of topology to? Dynamic systems seemed to be a fertile field. The switch was inspired by the work of the Dutch engineer Balthasar van der Pol. As we saw earlier, van der Pol was interested in how the frequencies in an electrical oscillator changed. He had discovered that some of the behavior was unpredictable.

Smale's work brings us back to a problem we stated earlier: the contradictory properties of strange attractors. Trajectories must converge since it is an attractor, but because of sensitivity to initial conditions they must also diverge. Smale showed that the problem could be resolved by a topological transformation in phase space.

It is obvious that, although trajectories diverge and follow completely different paths, they must eventually pass close to one another again. How can they do this? Smale showed that repeated stretching and folding of phase space can bring this about.

One of the simplest ways to demonstrate this is by using what is now called Smale's horseshoe. It is actually a series of transformations that create something that looks like a horseshoe. You start with a rectangle and stretch and squeeze it so it becomes a long bar. Then you take the end of the bar and bend it so it looks like a horseshoe. Finally put the horseshoe back in the original rectangle. You repeat this process by stretching and squeezing the rectangle until it is a bar again. Then bend the bar and put it back in the rectangle. Continue doing this indefinitely.

The space is obviously stretched in one dimension, squeezed in another, then folded. And this is exactly what happens to the

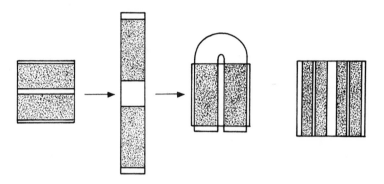

Smale's horseshoe. Space is stretched in one direction, squeezed in another, then folded. Process is repeated.

trajectories in phase space when we have a strange attractor. In particular, it resolves the enigma of the strange attractor that we mentioned above: how lines can diverge yet eventually get close to one another again. It also allows the attractor to be bound.

In short, the attractor takes nearby points and stretches them apart in a certain direction. This creates the divergence needed for unpredictability. Then the system "folds" these points so that points come together, causing a convergence. You can easily see this in the Lorenz attractor. Two trajectories move apart as one stays on the left wing and the other moves to the right wing of the butterfly. At the same time, trajectories from the wings are folded, and as a result they can come close together again.

Smale's horseshoe was an important breakthrough, part of a revolution that was taking place in our understanding of chaos. The discovery of strange attractors was another part; strange attractors, as we just saw, are fundamental in chaos theory and soon everyone was looking for new ones. Expectation was high that they would tell us something new about chaos.

But so far scientists had only seen strange attractors associated with equations. What about experimental evidence for them? That, we will see, was soon to come.

6

The Transition to Chaos

*T*he *Ruelle–Takens theory of turbulence was more appealing than Landau's theory. It required only three modes of vibration to* produce turbulence, whereas Landau's required an infinite number. But it did have a drawback: It required a strange attractor, and experimental verification would be needed to show that such an object existed.

How would you go about verifying the existence of a strange attractor? At the time Ruelle and Takens put forward their theory there was no way. At best you could show that experimental evidence favored one theory over the other, or that observations were inconsistent with one of the theories.

And this is basically what happened.

THE SWINNEY–GOLLUB EXPERIMENT

The first step in verifying the Ruelle–Takens theory was taken by Harry Swinney of City College of New York and Jerry Gollub of Haverford College. In 1977 Gollub took a sabbatical to come to New York to work with Swinney. Both were interested in fluid dynamic turbulence. Neither, however, knew anything about strange attractors, or the Ruelle and Takens theory. They had studied Landau's theory, and knew it was the accepted theory of turbulence, but they also knew it was not a proven theory, so they set out to prove it.

Early in the century a French hydrodynamicist, M. M. Couette, devised an apparatus for studying shear flow in fluids. His apparatus consisted of two cylinders, one inside the other, where water or fluid was introduced between the cylinders. One of the cylinders, usually the inner one, was then rotated at high speed, and the fluid was dragged along with it.

In 1923 Geoffrey Taylor of England sped up the inner cylinder and discovered an interesting instability that produced what looked like a pile of inner tubes, stacked one on top of the other. Later experimenters sped up the inner cylinder even more and showed that many other strange effects appeared: wavy vortices, twisted and braided vortices, and wavy spirals.

The pile of "inner tubes" that Taylor got was created by Landau's first transition. Looking at these tubes, it might appear as if little is going on, but if you place something in this region you will see that it moves around as the cylinder spins and also moves up and down.

Swinney and Gollub decided to take the experiment further. Their apparatus was small, only about a foot high and two inches across, with the space between the cylinders being about 1/8 inch.

A major difficulty in the early experiments was measuring the velocity of the fluid. Most experimenters used probes that disturbed the natural flow. By using a laser Swinney and Gollub were able to measure the velocity without disturbing the fluid.

Their technique centered on what is called the Doppler effect. You're likely familiar with this effect in relation to car horns or train whistles. As a car approaches blasting its horn, it has a higher pitch than when it is standing still, and when it passes, it has a lower pitch. This is due to the Doppler effect. The same phenomenon occurs with light: the frequency, or color, of the light changes as it approaches or recedes from you. If the light is stationary and you move towards or away from it you will also notice a change.

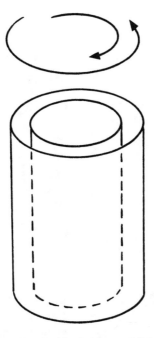

Apparatus used by Swinney and Golub.

To use the Doppler effect, Swinney and Gollub had to suspend tiny flakes of aluminum powder in the fluid. With the laser they were then able to measure the velocity of the flakes, which in turn gave the velocity of the fluid between the cylinders. Still, there was a problem. As the speed of the inner cylinder was increased many different velocities appeared, and they got a complicated signal—a mixture of waves of many different frequencies. They were able, however, to separate these frequencies through the use of a technique that had been invented by Joseph Fourier in 1807, a technique called Fourier analysis. Fourier analysis gives a "power spectrum"—a graph of the strength of each component frequency.

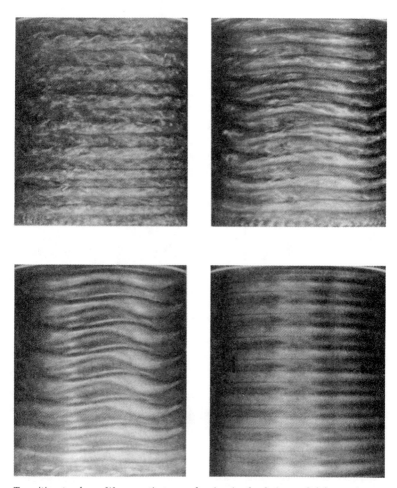

Transition to chaos. Wavy vortices seen forming in the Swinney–Golub experiment. (Jerry Golub)

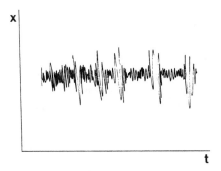

A "power spectrum" showing spikes.

If there is a major frequency present, a spike occurs in the power spectrum corresponding to that frequency. Considerable information can be obtained from this spectrum. If the signal is quasiperiodic, for example, several spikes occur. Finally a broad band of spikes indicates chaos.

Swinney and Gollub's apparatus was simple by today's standards; it sat on a desk, along with a few other small pieces of equipment. You don't see this very often in a lab today. Experimental equipment has become increasingly sophisticated and complex, with individual instruments costing tens and even hundreds of thousands of dollars. I often chuckle when I think back to when I was a graduate student in the 1960s. One of my teachers complained bitterly about how complicated and large equipment was getting. "You can't do a thing today without a huge crane," he said, shaking his head. I wonder what he would say today.

Swinney and Gollub were hoping to use the information they got from their equipment to verify Landau's theory. They set their apparatus running and soon saw the first transition predicted by Landau, and were pleased. To verify the theory completely, however, they had to observe many more transitions. They therefore began looking for the second, and to their surprise they didn't find it. Instead they got a broad band of spikes indicating chaos. They went through the experiment again and again, approaching the transition carefully, but each time they got the same thing. There wasn't a new frequency; the signal indicated chaos, or turbulence, immediately.

They weren't sure what to do, but they knew Landau's theory didn't predict what they were seeing. At this point they knew nothing of the Ruelle–Takens theory, so they couldn't compare their results to it.

Ruelle heard about this a short time later and made a trip to New York. He soon realized their data was consistent with his theory. Swinney and Gollub hadn't proved a strange attractor existed, or that Ruelle and Takens' theory was correct. However, their results, while inconsistent with Landau's theory, were generally consistent with Ruelle and Takens' theory, so it was a first step.

The major problem with the Swinney–Gollub experiment was that they couldn't look at all facets of the system at one time. In other words, they couldn't record the entire flow at a given instant over the system, only the flow at a particular point. Trying to obtain a strange attractor with such limited data was difficult, if not impossible.

The breakthrough that allowed the construction of a strange attractor from the data came a few years later (about 1980) from James Crutchfield, J. D. Farmer, Norman Packard, and Robert Shaw, working at the University of California at Santa Cruz. The technique was put on a firm mathematical foundation by Floris Takens about the same time.

The group at Santa Cruz used what would, by most standards, be considered a strange experimental setup—a dripping tap. It is well known that a tap drips regularly when it drips slowly. The water builds up on the circular rim of the tap, then drops off at regular intervals. If you turn the tap on a little more, however, something odd happens. The drips start coming at irregular intervals, in other words, at random. Crutchfield and his colleagues at Santa Cruz set up an experiment in which the drops hit a microphone and the time of each drop was recorded.

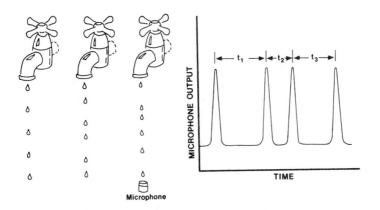

The dripping tap experiment. Plot of output from experiment is shown at right.

The technique developed by Takens and the Santa Cruz group required a measurement of the time interval between successive drops. These time intervals were, of course, random and appeared to be chaotic. By plotting them on a graph in a certain way (the details are too complicated to go into) Takens and his group were able to show that a strange attractor was present. Furthermore, evidence that the technique was valid came when they plotted the results from the Hénon attractor, a known strange attractor, in the same way. They got a plot that looked quite similar, giving additional evidence that there was, indeed, a strange attractor associated with the drops. Interestingly, when they made the drops come faster they got an more radical pattern that was not characteristic of a strange attractor.

There's more, however, that can be done with the above data. We'll illustrate with a chemical reaction that was discovered by B.P. Belousov and A.M. Zhabotinsky of the Soviet Union in the early 1960s. Zhabotinsky devoted his Ph.D. thesis to the study of the phenomenon.

The chemical reactions themselves are complicated, so we won't go into details. The important thing is that the chemical reactions oscillate back and forth (causing the concentration of a particular ion to oscillate), and under certain conditions the oscillations become chaotic. In 1980 J.C. Roux, A. Rossi, S. Rachelot, and Christian Vidal set up an experiment to monitor the change from periodic oscillations to chaos. Using a technique similar to that used in the dripping faucet experiment, they plotted the data from the oscillator in phase space. Their results showed conclusively that there was a strange attractor associated with the oscillating reaction. Later experiments by others verified the result.

YORKE, MAY, AND THE LOGISTIC MAPPING

We normally think of chaos as associated with physical systems—pendulums, vibrating metal beams, electrical systems, planets, and so on—but it is also an important phenomenon in biology, particularly in ecology where population studies play a central role. How, for example, does the population of rabbits in one year affect the population the next year, and in subsequent years?

James Yorke, a mathematician at the Interdisciplinary Institute at the University of Maryland, became interested in problems of this type in the early 1970s. His interest was sparked by Lorenz's paper on chaos. Chaos, he soon realized, was something new and exciting, something that he could make a contribution to. After studying the phenomenon in detail he wrote one of the classic papers of chaos, titled "Period Three Implies Chaos,"—a seemingly strange title but, as we'll see later, one with a message.

Yorke also gave the new science its name: chaos. Before this, scientists had a hard time communicating with one another. In fact, it was almost impossible for mathematicians, physicists, chemists, and biologists to talk to one another about the field. They all understood it differently.

Robert May followed in Yorke's footsteps. Like Yorke he became an ardent supporter of the new science. The two men were friends and frequently talked about chaos. May started out as a theoretical physicist in Australia. In 1971 he went to the Institute for Advanced Study at Princeton where he began to drift into biology, or more exactly, mathematical biology. Few mathematicians took an interest in biological problems at the time, so May had the field much to himself.

Over a period of years ecologists had accumulated a considerable amount of data. Many different kinds of insects, animals, fish, and so on had been studied to see how their populations varied from year to year. Biologists had even developed an equation that predicted the populations reasonably accurately; it was called the logistic equation or logistic mapping. It was a simple difference equation—a quadratic equation similar to the type most students solve in high school. It is hard, in fact, to imagine a simpler equation, but it seemed to work well, so biologists adopted and used it. It was so simple that it seemed unlikely that there would be anything mysterious or complex about it. In short, no one expected any surprises. As May soon found out, however, there were many surprises.

One of the main parts of the equation was a parameter that we'll call alpha. The equation itself is a difference equation, which means that when one value is substituted into it you get the next. For example, you could substitute the population of rabbits in a given year and you would get the population that would appear the next year. Continuing in this way you get a sequence of numbers. The best way to visualize what is going on is to plot the results on a graph; plot the input to the equation on the horizontal axis and the output on the vertical axis. For each possible input there is one and only one output. Making the plot you get an inverted parabola, like an upside-down bowl. The height of the parabola depends on the parameter alpha.

What we are really interested in is the long-term behavior of the system; in other words, what is going to happen to the population of rabbits over many years. And we can get this from the above graph. Begin by drawing a line up at 45 degrees

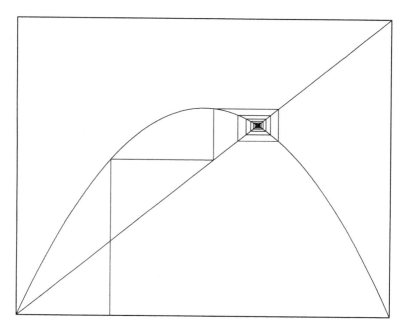

Logistic attractor showing path to attracting point. (The following diagrams show a sequence of increasing alpha.) Alpha is less than 3. Horizontal axis is input. Vertical axis is output. (Matthew Collier)

that intersects the parabola. Assume that in this case we have a parabola with a low value of the parameter alpha, say something less than 3 (see diagram). Now starting with an arbitrary value on the horizontal axis draw a vertical line up to the parabola then draw one over to the 45 degree line. Repeat this procedure all the way up to the top. When alpha is low the path up is simple. It quickly spirals into a point on the graph—a fixed point—corresponding to a single population.

Now, increase alpha slightly and do the same thing. You'll see that you get a fixed point on the parabola for a value of alpha from 0 to 3, but if alpha is greater than 3, say 3.2, things get a little more complicated. The path doesn't spiral onto the

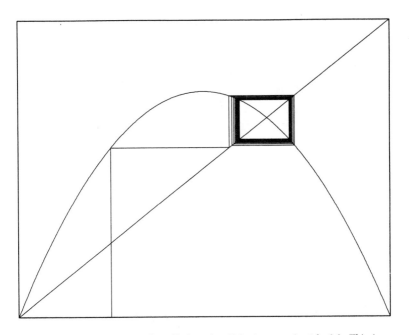

Logistic attractor showing path to limit cycle. Alpha is approximately 3.2. This is a period two cycle. (Matthew Collier)

parabola; instead it continues going around and around in something that looks like a limit cycle, only it's square.

What has happened is that the system has bifurcated; it has gone from steady state to periodic—to what is called a period two cycle. (The population fluctuates back and forth between two different populations.) If you increase alpha a little more you get a period four cycle. A little more gives a period eight cycle and so on. In each case you are going through a bifurcation. Soon everything breaks down and you get chaos. It sets in at an alpha of 3.57.

This was the equation that May was working with in relation to ecological populations. He explored the equation, examining the effects of different values of alpha, finding that if alpha

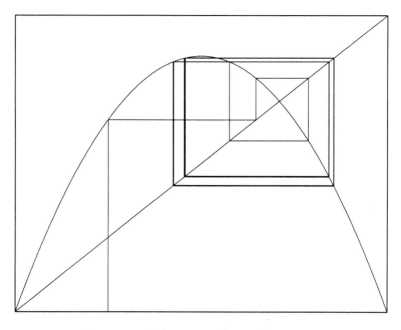

Logistic attractor. Alpha is greater than 3.2. (Matthew Collier)

was less than 1 the equation predicted that the population would decrease to zero. When alpha was less than 3 (but greater than 1) he found that the population settled into a constant value. Above 3, however, he found that what is frequently referred to as a "boom or bust" occurs. The population oscillates between two populations; one year it is high, the next it is low.

May was amazed with what he saw, and it spurred him to take a more global look at the predictions of the equation. He investigated hundreds of values of the parameter and plotted a graph of his results.

At an alpha of 3 he got a bifurcation to a period two population. As he continued to increase alpha he got another bifurcation at period four. This was a series of highs and lows over

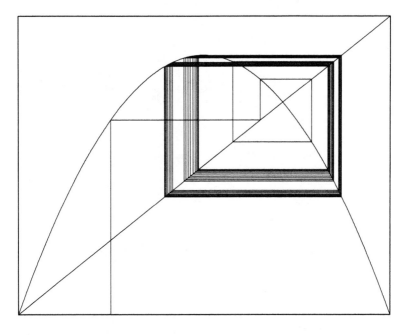

Logistic attractor. A period four cycle. (Matthew Collier)

4 years; the pattern repeated itself over a period of four years (for a population with a yearly reproduction cycle).

The bifurcations continued to get closer together. With another small increase the cycle doubles again and he got a period eight cycle, then a period sixteen and so on. Finally came chaos. When May plotted his results he got something that looked like the figure on page 120.

Gradually, as others joined in, the overall picture became clearer. After several bifurcations in which several doublings occur, the system suddenly becomes chaotic. But the strange thing is that within this region of chaos there are "windows" of order. These are regions where, even though alpha is still increasing, there is a periodic population, usually with a 3 or 7 year cycle.

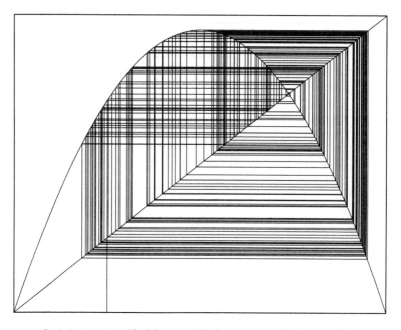

Logistic attractor with alpha over 3.57 showing chaos. (Matthew Collier)

They only last a while, however, with increasing alpha, then the bifurcations occur again with period doubling, and chaos takes over again.

One of the major things we see in the diagram is that the bifurcation structure appears to duplicate itself on smaller and smaller scales. If you look closely you see tiny copies of the larger bifurcations. This is the self-similarity we discussed earlier, and as we will see it is an important characteristic of chaos.

So far we have been dealing with equations—equations that presumably predict things in the real world. They tell us that ecological populations split in two, then four, and so on until finally chaos occurs and you cannot predict the population. But does this occur in nature?

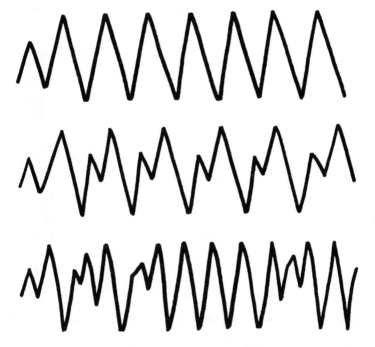

A simple representation of period one (top), period two (middle), and chaos (bottom).

Controlled studies have been made in the laboratory using small creatures such as blowflies and moths, and bifurcations do indeed occur, as the equation predicts. But there are always outside influences, and things aren't quite as cut and dry as they are in mathematics. Nevertheless, there is reasonably good agreement.

Yorke described many of these results in his paper "Period Three Implies Chaos." He proved that in any one-dimensional system, if a regular cycle of period three ever occurs, it will also display regular cycles of every other length, as well as chaotic cycles.

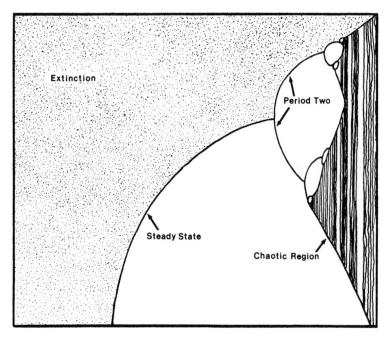

The period doubling cascade to chaos. "Windows" of order can be seen in the chaotic region to the right.

FEIGENBAUM

Chaos, it seemed, was much more complex than any one believed. There was much more to it than had been assumed, and as we will see, there was more to come.

What was needed now was somebody with patience and the right background, and this somebody appeared in the form of Mitchell Feigenbaum. Born and raised in New York, Feigenbaum excelled in school, particularly in mathematics, where he loved calculating and manipulating numbers—logarithms, trigonometric functions, and so on. His father was a chemist, his mother a school teacher.

Mitchell Feigenbaum. (Robert Reichert)

After graduating from City College in 1964 he went on to MIT where he got a Ph.D. in elementary particle physics in 1970. During the next few years he showed little of the creativity that would come later, but they were important years nevertheless; he was absorbing a lot of information that would become crucial later. He spent four years at Cornell and Virginia Polytechnic Institute with no publications. Then he went to Los Alamos National Laboratory in New Mexico.

Feigenbaum, like many creative people, was a dreamer, someone who seemed to have his head in the clouds much of the time. But in reality he was not dreaming idle dreams; he was thinking, thinking about chaos. His boss at Los Alamos had hired him to think. He knew talent when he saw it, and he knew that Feigenbaum had what it took. He expected great things from him. Strangely, though, for a period of several months after Feigenbaum arrived no one was quite sure what he was doing. What was he working on? It might seem strange that a person could have this much latitude—latitude to do almost anything he wanted, and be paid for it. University professors usually have considerable latitude to pursue whatever research they want, but they have lectures to teach. In most large companies and laboratories you usually work on a project that is assigned to you.

Feigenbaum's boss, however, knew that the gamble was worth it. He knew if he waited long enough something would come. And indeed it did. As it turned out, Feigenbaum had exactly what was needed for a major breakthrough in chaos. He was a theoretical physicist with a strong background in particle physics, an area quite distinct from chaos, but oddly enough, an area that had a connection to chaos.

The usual technique for solving particle interactions is called perturbation theory, a technique that involves drawing a lot of funny-looking little diagrams called Feynman diagrams, named for Nobel Laureate Richard Feynman who invented them. Feigenbaum spent several years making long, tedious calculations using these diagrams. Then one day he said to himself, "There's got to be something better than this."

And indeed, for him, there was. When he turned his attention to chaos at Los Alamos he never dreamed that his background in particle physics would be important. The equations in chaos were much simpler than the ones he had been dealing with, but the mathematics was no less tedious. However, he soon discovered that there was a whole new world out there waiting for him within a few simple little equations. He started doing what May and Lorenz had done earlier; he took one of the equations of chaos (the quadratic–logistic equation) and began look-ing into what happened as the parameter alpha was changed. He was sure that there was something there that May and others had not understood. But in his first attempt, he discovered little. In 1975, however, he travelled to a conference at Aspen, Colo-rado, and heard Steven Smale talking about chaos. Smale pointed out that the equations were simple, but he was sure there was a lot hidden in them that had not yet been discovered. The field was wide open, he emphasized, and important discov-eries were just around the corner.

Feigenbaum thought about it again. What had he missed? He went back to the logistic equation and began exploring the cascade of period doublings—the splittings—and he was soon amazed at the wealth of information that came pouring out.

Many people had looked into the splittings, and the chaos that resulted, but Feigenbaum looked further. What was signifi-cant in the way the bifurcations occurred? He began, as May had earlier, looking at them from a global point of view. Using his small pocket calculator (an HP-65), he began exploring the equations. You might wonder why he would use a small calcu-lator when he had access to some of the largest computers in the world, computers that could produce thousands, even tens of thousands, of numbers in a few minutes.

The reason, he said, is that he liked to play with numbers. Even when he was young he liked to play with numbers, see how they came about, how they changed under various circum-stances. It gave him a feel for what was going on. Having thou-sands of numbers handed to him on a sheet of paper that came from a computer wasn't the same; it didn't appeal to him. And indeed it was this "playing" that allowed him to make the break-through that he eventually made. It helped him understand what was going on. With large computers he probably would have missed it.

Using a tiny hand-held computer was tedious; it was slow, and he had to wait for each number as the computer ground away. To save time he finally began trying to estimate what the next number would be, and before long he was quite proficient at it. He was amazed, in fact, at how close he could come. He soon realized there was a distinct pattern to the way they came—they were closer and closer together. They were converging. He calculated the convergence and got 4.669. Was this number of any significance, he wondered? Perhaps a universal constant. He found it hard to believe, nevertheless he mentioned it to some of his colleagues. They thought it was just a coincidence.

What was particularly significant, however, was that a number such as this implied scaling. In other words, it implied that something similar was going on at a smaller scale. He decided to calculate the number a little more accurately, and for this he needed a large computer. Writing a simple program and running it, he got 4.66920160.

Then serendipity set in. On a hunch he began looking at another equation known to produce chaos, an entirely different equation based on trigonometric functions. Again bifurcation came as splittings occurred, in the same way as they did with the logistic quadratic equation. Things looked so similar he decided to calculate the convergence of the sequence, not expecting it to be related in any way to the convergence of the quadratic equation. After all, they were completely different equations that didn't even resemble one another. Their only connection was that they both gave rise to chaos.

To his amazement Feigenbaum got the same number: 4.66920160. It was impossible. How could it happen? The equations were completely different. He went carefully through the calculations, taking both to many significant figures. It wasn't a coincidence; the two numbers were the same.

The number he got was obviously a universal constant, in the same way that pi and *e* are universal constants. Feigenbaum realized that there had to be a reason for this number—an underlying theory. He began looking into it, and this is where his background in particle physics was helpful. The number implied scaling, and scaling reminded him of a procedure called renormalization in particle physics, a technique that had been suggested as early as the 1930s. Quantum electrodynamics, the theory of the interaction of electrons and photons (particles of light), had been at an impasse because the perturbation expansion gave infinities that completely destroyed any meaningful output from the theory. Victor Weisskopf of MIT suggested that the infinities could be eliminated by absorbing them in a redefinition of mass and charge. In other words, the mass and charge of the particles could be rescaled. Hans Bethe, Julian Schwinger, and others showed a few years later that it worked, and soon renormalization was an integral part of field theory.

Ken Wilson of Cornell became interested in the idea many years later in regard to a different problem. Wilson, like Feigenbaum, started slowly, then suddenly blossomed over a period of a few years, years that produced several epic-making papers. He did his Ph.D. thesis under Nobel Laureate Murray Gell-Mann. When he graduated he went to Harvard, then to Cornell. At Cornell he became interested in what is called the renormalization group. He developed the renormalization group technique and showed it could be used fruitfully in particle interaction.

Wilson's renormalization method could be used not only with particle interactions but also in the study of phase transitions—transitions that take place when a substance changes from a gas to a liquid, or a liquid to a solid. Feigenbaum saw immediately that the transition from steady state through several splittings to chaos was similar to phase transitions. Also, scaling was involved in Wilson's theory, and scaling therefore had to be involved in the transition to chaos.

What is renormalization? First of all, if scaling is present we can go to smaller scales and get exactly the same result. In a sense we are looking at the system with a microscope of increasing power. If you take the limit of such a process you get a stability that is not otherwise present. In short, in the renormalized system, the self-similarity is exact, not approximate as it usually is. So renormalization gives stability and exactness.

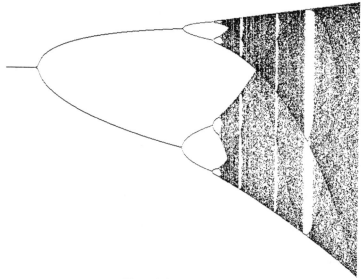

The period doubling cascade.

If you look at the branching that occurs during the cascade of splittings you see that you can take a limit, and with the renormalization group, this limit is exact, not approximate. With his theory Feigenbaum was able to explain the significance of the number 4.66920160.

Strangely, when Feigenbaum wrote up his paper and sent it off to be published it was rejected, and he didn't have the option that Ruelle had when his paper was rejected. Still, he knew the result was important, so he went on the lecture circuit and began spreading the word. Soon there were hundreds of requests for reprints of his papers—the papers he had been unable to publish.

OTHER ROUTES TO CHAOS

With our emphasis on the cascade of period doubling route to chaos it may seem that it is the only one. But it isn't. Earlier we saw that Ruelle discovered a different way, a way that required only three different modes of vibration. His route is usually called the quasiperiodic route. The first transition occurs, then the second. When the frequencies of the two are not commensurate you get quasiperiodic motion. This motion is assumed to take place on the surface of a torus. Chaotic motion occurs when the quasiperiodic torus breaks up as the parameter is varied. The torus is replaced with a strange attractor.

Another route to chaos is referred to as intermittency. In this case chaotic spikes occur in the spectrum, and as alpha is varied they become more frequent and finally take over the system.

An understanding of the transition to chaos was an important breakthrough. We saw that the logistic equation, an equation that gives predictions of ecological population growth played a key role. Through a detailed study of this equation scientists saw that a sequence of bifurcations led to chaos. Furthermore, they discovered a universal constant related to the cascade of period doublings.

7

Fractals

*We discussed fractals briefly in the last chapter, defining them to
be geometrical forms that appear the same on all scales. Fractals*
and chaos developed independently, and at first appeared to be
unrelated. But slowly, as a better understanding of them
emerged, scientists began to realize they were closely related.
Fractals eventually supplied a new and useful language to
express chaos. It is important, therefore, to take a closer look at
them.

COASTLINES AND CURVES

Looking around you see many examples of fractals. Nature,
in fact, is full of them. A good example is a coastline; if you
look at the western coastline of the United States on a map, it
appears relatively smooth. If you flew over it in an airplane,
however, you would be surprised at how jagged it was as compared to the map. The reason, of course, is that small details
cannot be shown in a map because the scale is too large.

But even from an airplane you're not seeing a lot of the
detail. If you drove along the shoreline you would see many

things you didn't see from the airplane. In fact, the closer you got to the shoreline, the more bays, beaches, and so on you would see. In a true fractal (a shoreline is not a true fractal, only a good approximation) this detail would continue indefinitely as you got closer and closer.

Coastlines are not the only fractals in nature. The boundaries between many countries are fractal. Rivers are fractal; if you follow a river upstream you see it is composed of tributaries, and the tributaries, in turn, are composed of smaller tributaries. Of course this process doesn't go on indefinitely, so rivers are only approximately fractal. The same goes for trees; if you follow a limb up you see branches coming off it, and if you follow one of them, you see branches coming off it. Clouds and the jagged tops of mountains are also fractals.

One of the most important properties of fractals can be seen by going back to our coastline. Assume for simplicity that we are dealing with an island, and we want to measure the distance around the island (its perimeter). To get an approximate value we use a map. The scale of any map is usually given in the lower right hand corner. Assume the unit is 100 meters, and we take a ruler and measure the perimeter of the island in terms of this unit, and get 4783 meters. Looking at the figure, however, it's easy to see that we have skipped over a lot of bays and peninsulas, so it's only an approximate measure. If we shorten the measuring rod to, say, one meter, we'll obviously get a number that is much closer to the true value. This time we'll walk around the island with our measuring rod, and we get 7842 meters.

As a third step let's shorten our measuring rod to one centimeter; in this case we get 11,345 meters. You may be asking at this point: Will this sequence of numbers converge to the exact length of the shoreline? In a true fractal, if we continued this process indefinitely, we would get a larger number each time, and in the limit of an infinitesimally small measuring rod we would get an infinite length for the shoreline.

How could an island of finite area have an infinitely long coastline? As strange as it seems, this is one of the major prop-

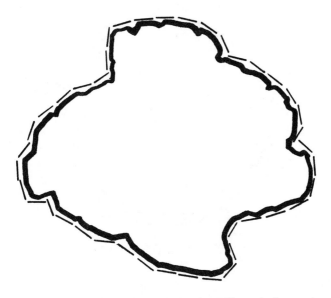

An island showing measuring sticks around the edge. Different-sized measuring sticks will give a different perimeter.

erties of a fractal, and it's a property smooth curves don't have. This can easily be seen by measuring the perimeter of a circle in the same way. We'll start with a relatively large measuring rod that takes, say, eight lengths to go around it; this would be like inscribing a hexagon in it.

Taking a smaller measuring rod would give more sides, and gives a longer length which would approximate the true length better. If we took an even smaller rod we would get a larger number again. Unlike the previous case, however, in the limit of an infinitely short rod we do not get infinity; we get 3.1415 times the diameter (the usual formula for perimeter), and this is a finite number.

What are the implications of this? First of all, it means that the geometry we use for smooth curves, namely Euclidean

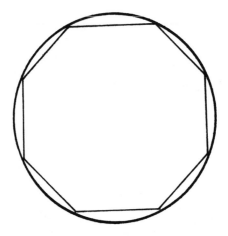

Hexagon inscribed inside circle.

geometry, won't work when applied to fractals; we'll need a new geometry.

MANDELBROT

Shortly after Benoit Mandelbrot came to IBM in New York in the 1950s he noticed a new phenomenon. He was surprised that it was so prevalent, and no one else had studied it. In fact, it didn't even have a name. Over a period of years the foundations of a new science based on this phenomenon began to form in his mind.

Born in Warsaw to a Lithuanian Jewish family in 1924, Mandelbrot's early education was sporadic and incomplete. When he was 12 his family moved to Paris where he met his Uncle Szolem, who was a mathematician. Szolem introduced him to mathematics, and encouraged him to take it up, but when Mandelbrot was ready to go to the university, World War II broke

Benoit Mandelbrot.

out and his family was forced to flee from Paris. They went to the south of France but weren't able to escape the Nazis, and for several years lived under their rule.

In 1945, after the war, Mandelbrot took the entrance exams for the two most prestigious schools in France: École Normale and École Polytechnique. He was not adequately prepared for the exams and had to rely on his natural ability to reason, and on a peculiar ability he had for using geometric figures to assist him in his thinking. As might be expected, he did poorly on the physics and chemistry sections of the exam, but did so well on the mathematics part that he was accepted.

He selected École Polytechnique and over the next few years developed a deep understanding and love for mathematics. His uncle had recognized his ability and encouraged him, but strangely he had also given him a stern warning: avoid geometry. In particular, avoid the use of pictures, which he said could be misleading. Solve problems analytically. Mandelbrot took the advice seriously but soon realized it was not the best advice for him. He had an uncanny ability for visualizing problems geometrically. It was second nature to him, and something he needed in his thinking.

In the late 1950s Mandelbrot came to the United States, settling at IBM's Thomas J. Watson Research Center. He was young and full of ideas, and the position turned out to be an ideal one for him; it gave him considerable latitude to develop and use his talents.

During his first few years at IBM he worked in many different areas. To some he appeared to be jumping from problem to problem, working for a few months on a problem in economics, then abandoning it for a problem on transmission wires. What few, if any, noticed was that there was a common thread throughout the problems he was working on.

One of the problems that attracted his attention was that of noise in the telephone lines used to transmit information between computers. The noise was sporadic, but occasionally it came in bursts and wiped out some of the information being carried along the line. Engineers tried several methods to get around it, but had become frustrated in their attempts.

Mandelbrot took a different approach. He tried to understand the noise by setting up a mathematical model and using this model to see how the noise could be controlled. He began by studying the structure of the bursts, something few others had done. Noise, it was well known, was random, and it seemed unlikely anything could be gained by studying its structure. But Mandelbrot saw something that others had not seen: when he magnified the bursts, he saw that they were self-similar.

The sequence of bursts brought to mind something he had studied years earlier—a set discovered by the German mathe-

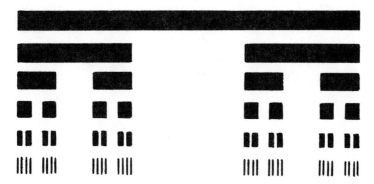

The Cantor Set.

matician, Georg Cantor. Cantor had taken a line, deleted the center third, then deleted the center third of the remaining pieces, and so on all the way to infinity. What he got was a sequence of points that had zero length, a sequence that is now called the Cantor set. To Mandelbrot the sequence of points looked exactly like the plot of the bursts of noise he was seeing in telephone lines.

Mandelbrot soon recognized the phenomenon in other areas. It had to have a name, and in 1975 he coined the term "fractal." Convinced the phenomenon was universal and of considerable importance he began writing and publishing papers on it, but strangely, for the first few years they were ignored by the scientific community. Disillusioned and somewhat disgusted, he put everything he knew about the subject into a book entitled *Fractals: Form, Chance and Dimension.* This was followed a few years later by an expanded and refined version titled, *The Fractal Geometry of Nature.* The books were extremely successful. No other book on mathematics, in fact, has approached their sales. One of the major attractions of the books was the amazing graphics in them—graphics that could easily be simulated on a computer—and because of this they were bought in large numbers by computer enthusiasts.

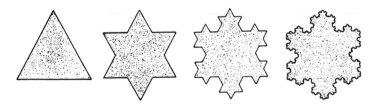

The Koch snowflake.

Still, Mandelbrot's ideas were not instantly accepted by the scientific community; it took many years to convince scientists of their importance.

MATHEMATICAL FRACTALS

As we have seen, nature is full of fractals. But there is another way of creating fractals: by defining them mathematically. One of the first to do this was the Swedish mathematician Helge von Koch. In 1904 he described what we now refer to it as Koch's snowflake. To construct it we start with an equilateral triangle.

As in the case of the Cantor set, we cut out the middle third of each side, but this time we replace it with a triangle similar

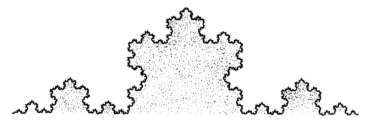

Closeup of the Koch snowflake pattern.

to the original one, but smaller. We continue doing this to each of the new sides until finally, in the limit, the perimeter becomes infinite. Again we have a mathematical monstrosity: an object with a finite area but infinite perimeter, and it is, of course, a fractal.

There are many other curves like this that can be constructed. One is called the Sierpinski gasket. As with the Koch snowflake, you start with a triangle. Mark the midpoints of each side and form a triangle by joining them. Delete this triangle and mark the center points of the sides of each of the remaining triangles; form new triangles using these points, and delete them. As you continue, the triangles will get smaller and smaller, and you will have triangles nested within triangles. And just as the Cantor set has zero length, this set, in the limit, has zero area, and is a fractal.

You can do the same thing using squares instead of triangles. In this case you start with a square. Treating each side like a Cantor set, you divide it into thirds, mark the square at the center, and delete it. Continuing this ad infinitum gives another object called a Sierpinski carpet that has zero area. This object can be extended to three dimensions in which case you get what is called a Menger sponge, an object with zero volume.

A NEW GEOMETRY

As we saw earlier we can't use Euclidean geometry when dealing with fractals. We need a new geometry. In formulating this geometry one of the first things we look at is "dimension." In Euclidean geometry a point has dimension zero, a line dimension one, a surface dimension two, and a volume dimension three. What is the dimension of a fractal? It might seem, at first, that a coastline, or the boundary of a Koch snowflake, would be one-dimensional. You're measuring the distance along a line, which in Euclidean geometry is one-dimensional. But a fractal "wiggles" (e.g., the irregularity of a coastline) and these wiggles

The Sierpinski carpet.

take up area, so in another sense it is two-dimensional, yet it's not completely two-dimensional. In fact, it is defined as having a dimension between one and two, in other words, a fractional dimension. The surface of the Koch snowflake, for example, has a dimension of 1.2618.

This might not seem to make sense. How can you have something with a dimension between one and two? If you think about it for a moment, however, it does make sense. A fractal such as a coastline or Koch snowflake obviously has to have a dimension greater than one, otherwise we couldn't distinguish it from an ordinary smooth curve, yet it can't have dimension two, or it would be a surface.

Felix Hausdorff, a German mathematician, introduced the idea of fractional dimensions in 1919, but it was Mandelbrot who brought it forcefully to our attention. After introducing the idea of a fractal he reintroduced the concept of a fractional dimension. It applies not only to jiggly lines but also to higher dimensions. The fractal dimension of the Sierpinski carpet, for example, is 1.8928, and the fractal dimension of the Menger sponge is 2.727.

What does this number tell us? Looking at the fractal dimension of the Koch snowflake, namely 1.2618, we see it is 26 percent greater than that of a line. The .26 gives us a measure of how "jiggly" the line is. In this case it has moderate jiggle. If it was 1.973, for example, it would almost be "area-filling," almost two-dimensional. The Sierpinski carpet has a dimension 2.727 indicating it is relatively close to three dimensions.

Earlier we talked about attractors, and as you might expect, they are also related to fractals. Ordinary attractors are associated with simple curves, but strange attractors are associated with complex curves, curves that are fractal. Thus strange attractors are fractal. You can, in fact, calculate the dimension of all the strange attractors we discussed earlier, and in each case you will see they have fractional dimensions. The Lorenz attractor has a dimension of 2.06, and the Hénon attractor has a dimension of 1.26.

OLBER'S PARADOX

A few years after Mandelbrot began looking at fractals he realized the concept was important in astronomy. A paradox had arisen many years earlier. It had been named after the nineteenth-century amateur astronomer Wilhelm Olbers, but actually had been known for many years before Olbers studied it. One of the first to consider it was Kepler. At the time, one of the major problems in astronomy was: Is the universe finite or infinite? If it was infinite, it should have an infinite number of stars in it, and according to Kepler the night sky should be bright—but it wasn't.

Stated in such a simple way, the paradox is hard to take seriously. But over the years it became evident that it was a serious problem. To understand it more fully let's divide the universe into concentric spheres centered on the Earth, where the distance between the shells is constant. Also, assume the stars are uniformly distributed, and on the average they are approximately the same brightness. As we look out through the con-

Olber's paradox. Shells around the observer are of equal size.

centric shells, the more distant stars will appear dimmer because the intensity of light drops off with distance. But as we go outward the volume within each shell also increases, so there are more stars in it. In fact, the increase in number of stars exactly compensates for the decrease in intensity. In more technical language, the light intensity drops off as the square of the distance, but the number in each shell increases as the square of the distance. What this means is that each shell contributes an equal amount of light, and if there are an infinite number of shells, there will be an infinite amount of light. And this in turn means the night sky can't be dark.

Edmund Halley of Halley Comet fame was one of the first to bring the paradox to the attention of the scientific world; he published two papers on it in 1720. The first resolution of the paradox came in 1908 when C.V. I. Charlier showed mathematically that if there was a hierarchy of clustering in the universe the night sky need not be bright. A hierarchy of clustering implies that the universe is made up of clusters, clusters of clusters, ad infinitum. Such a hierarchy was not known at the time and Charlier's solution was soon forgotten.

With the discovery of the expansion of the universe astronomers realized that distant objects would be dimmed because of their redshift; objects in the outer parts of the universe would thus not contribute as much light as expected. Edward Harrison of the University of Massachusetts, however, showed that this doesn't completely resolve the problem, and he has given a resolution that is now generally accepted. He showed that the universe is not old enough, and therefore not large enough, to have a bright night sky. Most of the light needed to give a bright night sky would, in theory, have to come from distant parts of the universe—more than 10^{24} light years away, and the universe is not that large. As we look out into the universe we are looking back in time, and see younger and younger stars, until at about 10^{10} light years there are no more stars.

Mandelbrot was not satisfied with this solution. Looking back at Charlier's solution (if the universe is self-similar—the same at all scales—the paradox is resolved) he saw that what it boiled down to was: the paradox is resolved if the universe is a fractal.

Is the universe a fractal? Several studies have been made recently, and it doesn't appear to be an exact fractal, but it is close. We'll have more to say about this later.

JULIA SETS

Another relationship between fractals and chaos comes in the Feigenbaum sequence of numbers that leads to chaos. Michael Barnsley, an English mathematician, learned about this sequence in 1979 and became intrigued by it. What was its origin? How could it be explained? He was sure there were things about it that had not yet been uncovered, and after a brief study he was pleased to find that he had found something nobody else had noticed. He had an explanation of the sequence.

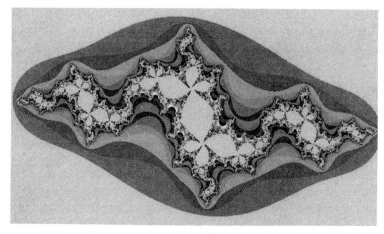

One of the Julia set. (George Irwin)

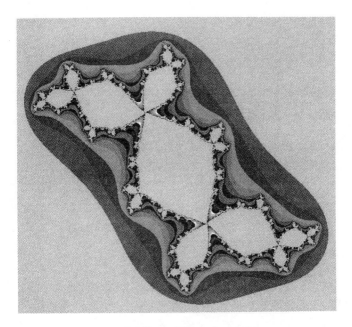

A Julia set. (George Irwin)

He wrote a paper and submitted it for publication. The editor of the journal, however, wrote back telling Barnsley that the mathematical sequence he had discovered was well-known. It had been discovered in 1918 by the French mathematician Gaston Julia, and was known as the Julia set. Barnsley was disappointed. The discovery, however, was important nevertheless: It was the first link between chaos and fractals.

It's difficult to define the Julia set without using mathematics, but I'll try to keep it to a minimum. The Julia equations involve complex numbers, so let's begin with what is called the complex plane. As you likely know, the set of real numbers can be plotted along a line. But there are numbers that can't be plotted on this line. To see where they come from consider the operation of taking a square root. The square root of a number is

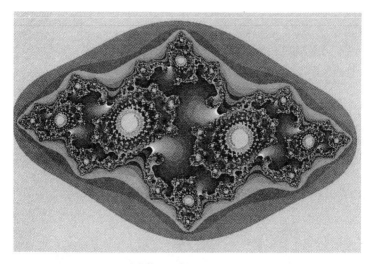

A Julia set. (George Irwin)

another number that when multiplied by itself gives the first number; the square root of 4 is 2, and the square root of 9 is 3. But what about the square root of –4? What two numbers, when multiplied together give –4? Mathematicians invented a new type of number to account for such cases; they are called imaginary numbers. If we define the square root of –1 to be i, and call this the imaginary unit, then the square root of –4 is $2i$, and we can plot this number just as we can plot the real numbers; the only difference is that we plot it along a different axis.

A complex number has both a real and an imaginary part; the real part, which we will call "a" is a real number, and the imaginary part, which we will call "b" is an imaginary number. We write it as $a + ib$. And just as real numbers can be plotted along a line, complex numbers can be plotted in the complex plane where one of the axes (the horizontal) is the real axis, or axis of real numbers, and the other one (the vertical) is the imaginary axis. Thus we have complex numbers such as $3 + 3i$ or $5 + 2i$. We usually refer to such numbers as z.

Now, back to the Julia set. How do we get it? We start by selecting a given complex number (i.e., a number such as 3 + 3i); for convenience let's call it z. We multiply it by itself; in other words we square it (the square is written z^2). We then add a constant number to it; call this constant c. Since we are dealing with complex numbers c is a complex constant. Mathematically we write these two operations as $z \rightarrow z^2 + c$. Now take the number you get after squaring and adding c and square it again and add c again; in other words do the same operation again. You continue doing this indefinitely for the z you have selected. The number that you finally get may be finite, or infinite. You then go to a different initial z and do the same thing again, and again you will get either a finite or infinite result. Do this for a large number of z's. The Julia set is the boundary between the infinite results and the finite ones.

Several examples of Julia sets are shown in the figures. As a special case consider $c = 0$. It's relatively easy to see that in this case that we just get the squares z, z^2, z^4, z^8 and so on. If z is less than 1 this sequence will remain bound; if it is greater than 1 it will go to infinity. The Julia set is therefore a circle of radius 1 around the origin. In terms of what we learned earlier we can say that the region inside the unit circle is the basin of attraction for a point attractor at the origin.

If you now change c slightly from 0, you get a distorted, but connected, region where the attractor is no longer at the origin. If you continue making c larger you find that this simple region breaks up into several connected islands, each of which has an attractor in the center. Finally, however, for larger c, the islands become disconnected.

THE MOST COMPLEX OBJECT IN MATHEMATICS

Mandelbrot became intrigued with Julia sets in the late 1970s. In 1979 he took a leave of absence from IBM and went to Harvard as a visiting professor. Harvard had a large VAX

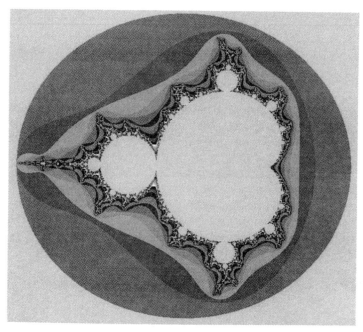

The Mandelbrot set. The Mandelbrot snowman. (George Irwin)

computer and Mandelbrot and an assistant began using it to explore Julia sets. His ideas on fractals were now firmly established and he wanted to see exactly how Julia sets were related. He had taken a course from Julia back in France and had become familiar with Julia sets when he was about 20, but had not thought about them much since.

After exploring and studying several Julia sets on the new VAX, the novelty wore off, and Mandelbrot started to examine similar, but more complicated mappings to see what they gave. He was sure that simple equations like the one that defined the Julia sets would give little new information. But after exploring several complicated equations and getting nowhere he came

A closeup of a section of the Mandelbrot set showing a snowman. (George Irwin)

back to a simple one, similar to Julia's. In fact, it's the same equation, but looked at from a different point of view.

Again we have $z \rightarrow z^2 + c$ where we start with an initial z and look at the ensuing sequence. This time, however, we always start with $z = 0$ and vary c. For some values of c the sequence will go to infinity, for others it will remain finite. Again we're interested in the boundary between the two regions. Mandelbrot plotted them up and got a strange-looking figure, something that looked like a snowman lying on its side, with a lot of twigs and balls attached to it.

He decided to magnify the boundary so he could get a closer look at it, but when he did, it got "messy." In other

A closeup of a section of the Mandelbrot set. (George Irwin)

words, it became diffuse and he couldn't make things out, so Mandelbrot rushed back to IBM with its larger computers to see if he could get a clearer picture. And indeed he did, and was amazed by what he saw. The picture was incredibly complex.

What did it mean? Mandelbrot began exploring the new figure. As he magnified its borders more and more, all kinds of strange figures emerged. Mandelbrot saw immediately that the figure was self-similar and the object was a fractal. But it wasn't a fractal in the same way the Koch snowflake is. In other words, it wasn't exactly self-similar. As he continued to magnify it, all kinds of weird things emerged: seahorses, scrolls, coils of all types. It was truly an amazing object, and most amazing was that when he looked at it at higher magnifications he discovered

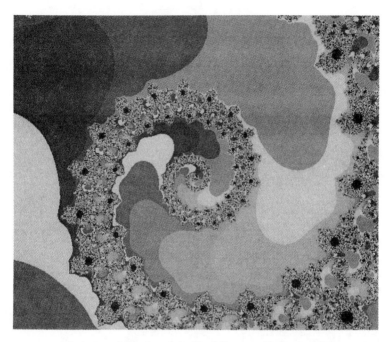

A closeup of a section of the Mandelbrot set. (George Irwin)

replicas of the whole object on a smaller scale—tiny snowmen tucked into the pattern.

What is really surprising is that such a complex object is generated by such a simple equation. About 10 lines of computer programming is all that is required to generate it. If you tried to store this much information in the computer memory, it would probably overcome the computer.

As Mandelbrot examined the figure further he discovered to his amazement that all the Julia sets were embedded in it. Different sets at different points. In particular, if you considered a path beginning in the interior of the figure, which is now called the Mandelbrot set, and extended to the exterior, you would find the Julia sets on the inside are connected, but as you get to the boundary they separate, almost as if they are exploding.

In 1991 the Japanese mathematician Mitsuhiro Shishikura showed that the fractal dimension of the Mandelbrot set was 2. It might seem strange that a fractal would have an integral dimension, but in this case it is significant. Remember that a boundary is one-dimensional, with a little extra depending on how complicated the "jiggling" of the boundary is. The fractal dimension can be anywhere between 1 and 2 depending on the complexity. In short, the maximum is 2. This means that the Mandelbrot set is the most complex fractal of its type; it has the maximum fractal dimension, and this is why most people now refer to it as the most complex mathematical object ever discovered.

FAKE FRACTALS AND COMPUTERS

If you flip through one of the recent books on fractals one of the first things you see are figures that look like something out of a science fiction movie. There are planets, surfaces of moons and even alien Earthlike landscapes, but they're not real. Even the ones that resemble the surface of the moon are fakes. To store all the information needed to graph even a small section of the moon's surface in a computer takes an incredible amount of storage. Computer graphics coupled with fractal techniques, however, can now give amazing fakes. They are now used extensively in movies and books.

Why use these rather than real pictures? The reason is that it takes only a short computer program to generate them, whereas it takes a tremendous amount of memory to store the pictures directly.

The technique for generating them centers around repetition, and as you no doubt know the computer is particularly good at repetitive processes. Millions of repetitions per minute are now common. But where do fractals come in? Fractals break up into smaller and smaller copies of themselves. Using these small copies we can "build" scenes on a graphic output using a computer. The figures you see in these books are made this way.

NEWTON'S METHOD

Every once in while an old method is shown to have something within it that no one expected. Such was the case with Newton's method for solving polynomial equations. With Newton's name attached to it, it is obviously an old method.

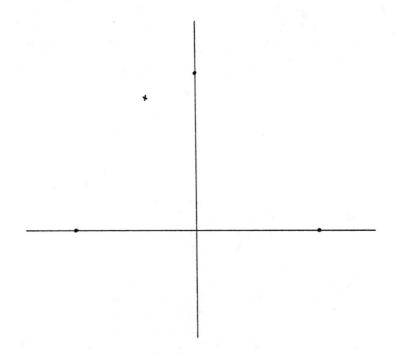

Newton's method of solving polynomial equations. The three solutions are shown as dots. The trial solution is shown as an x.

Pattern from Newton's method. (George Irwin)

Nevertheless, it is a useful and efficient method of solving equations, a method that centers around iteration. The fact that we are dealing with iteration may give you a clue, since Julia and Mandelbrot sets are generated by iteration.

Any undergraduate in mathematics is taught the method. You start with a guess. The guess leads to a better guess and so on. By continuing the process you finally get a solution.

To see some of the details let's go again to the complex plane. A third degree polynomial is going to have three solutions in this plane. We represent them by three dots. The object is to find the three dots (solutions) by use of Newton's method.

You start with a guess. If it is relatively close to one of the three solutions, Newton's iteration will take you rapidly to it. But what happens if you select a point that is exactly halfway

Pattern from Newton's method. (George Irwin)

between two solutions? John Hubbard, an American mathematician, became interested in this question in the 1970s. Using a computer, Hubbard began exploring this boundary. Note that there is a similarity here to the sets discussed earlier: We have an attractor (the solution) and a basin of attraction around it. You might think the boundary between the various basins would be simple—perhaps just a line. But Hubbard found differently. He found that it was extremely complex, and of particular importance; it had a fractal structure. If you start very close to the boundary the guess might give a series of points that bounce around chaotically many times before converging toward one of the solutions. In some cases—right on the boundary—the point will fall into a cycle that never converges; it just repeats itself over and over. Hubbard had the computer plot his attempts in color and he got some amazingly beautiful graphs.

In this chapter we have seen that fractals come in many shapes and forms, and in recent years with the development of computers there has been a tremendous interest in them. For us they are important because they are intricately related to chaos. Strange attractors, which are at the heart of chaos, are fractal.

This concludes our look at the basic nature of chaos. With this background we can now consider chaos on a grander scale, namely chaos in the universe.

8

Chaos in the Solar System—Introduction

*I*n our search for chaos in the universe, the logical place to start is the solar system. There were indications early on that chaos might play an important role in the orbits of the planets. As we saw in an earlier chapter, Poincaré encountered chaos near the beginning of the century while looking into the long-term stability of the solar system. At the time the solution of the two-body problem was well-known; it was relatively easy and had an analytical solution. When Poincaré added a third body, however, he was surprised how monstrously difficult the problem became. He showed that three astronomical bodies under mutual gravitational attraction had no analytic solution. But Poincaré wasn't one to give up easily. Instead of pursuing the straight forward approach he tried a round-about one: he investigated the orbits qualitatively by plotting them in phase space, then he examined a slice through the trajectories. And in the process he discovered chaos.

Many people were discouraged by Poincaré's discovery. It seemed to imply that it wasn't worth pursuing the problem; nothing would be gained. For years few people worked on it, but the few who did carried the torch forward. Even though the problem couldn't be solved directly, it could be approximated closely using perturbation techniques. Over the years perturbation techniques improved significantly, until finally astronomers could, in theory, get answers to any degree of accuracy.

Still, there was a problem: Long hours of tedious calculations were needed to get high accuracy, and few wanted to indulge themselves. Months and even years of long routine calculations were required. And after all the work the results were sometimes questionable because of the approximations that were used.

THE KAM THEOREM

An important breakthrough in our understanding of the stability of the solar system came in the 1950s and early 1960s with the formulation of the KAM theorem, named for its Russian discoverers, Andrei Kolmogorov, Vladimir Arnold, and Jurgen Moser. The theorem applied only to conservative or nondissipative systems (systems that don't lose energy), but that included almost everything of interest to astronomers in the solar system. Kolmogorov took the initial step in the early 1950s; he considered the stability of a hypothetical system of several planets that was under a small perturbation. He did not obtain a solution for the problem, but outlined a plan of attack, and the details were carried out by his student Vladimir Arnold in 1961. Moser extended their results a few years later.

The phenomena of resonance between orbital periods plays an important role in the theorem, so it's best to begin with it. Any two associated periodic motions can be in resonance; a pail swung around in a loop by someone riding a ferris wheel is a simple, if somewhat silly, example. If the person swings the pail around ten times while the ferris wheel goes around once, the two are in a 10:1 resonance. As you might expect, resonances of this type occur in the solar system. Two moons in orbit around a planet, for example, can be in resonance; if one goes around twice while the other goes around once they are in a 2:1 resonance. Furthermore, both may be in resonance with the planet's orbital period. Resonances such as this play an important role in chaos.

Kolmogorov began his study by considering a simple system of planets for which a solution existed. He then asked himself what would happen if a small perturbation was applied to this system. Small perturbations are common throughout the solar system; the Earth, for example, is perturbed by several objects. Although its orbit is determined primarily by the strong gravitational pull of the Sun, our Moon, and even distant Jupiter perturb it. Fortunately their perturbations are small, and unlikely to cause chaos in the near future.

The natural place to study such a problem is phase space, and Kolmogorov and Arnold quickly turned to it. In phase space we know that when a system has two associated periods the "phase point" is confined to the surface of a torus. One of the periods is associated with the short distance around the torus (C_1 in the diagram) and the other with the long distance

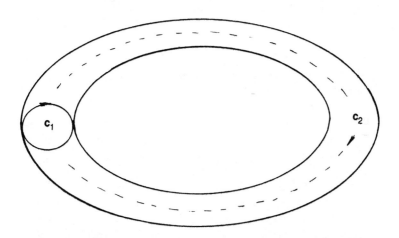

Two radii associated with a torus: short radius C_1 and long radius C_2.

(C_2 in the diagram). The ratio of these periods is critical to the behavior of the system. If the ratio is a rational number (can be written as p/q where p and q are integers) then the motion of the system will be periodic, and the state point will eventually return to the same point on the torus. If, on the other hand, it is irrational (such as $\sqrt{2}$) the motion will be quasiperiodic, and the orbit will wind over the entire torus, and never come back to the same point. In theory, though, it can approach any point as closely as we want.

Kolmogorov and Arnold found that periodic orbits become unstable under perturbation, whereas quasiperiodic orbits remain stable. This was the essence of their theorem. More formally it can be stated as: *If you start with a simple linear system for which a solution exists and add a small perturbation, the system will remain qualitatively the same.* The theorem has important implications for the long-term stability of the solar system. Moser, in fact, showed that a system of planets such as the solar system would remain stable only if planetary masses, orbital eccentricities (elongations of the orbit), and the inclinations of the spin axes were small. Our solar system, unfortunately, does not rigorously satisfy these requirements.

To see the significance of the KAM theorem let's look closer at a planet in phase space. Depending on its speed it can move in one of several different orbits. In phase space we therefore have a series of concentric tori, one inside the other. At the center is a periodic orbit; the tori surrounding it are associated with secondary periods. If you take a Poincaré section through these tori you get a series of concentric circles, as shown in the figure.

The motion is confined to one or other of these circles, some of which are periodic, others quasiperiodic. We assume that initially there is no perturbation (this is the stage shown). Now assume a small perturbation is introduced. Some of the tori will be destroyed, and as the perturbation is increased, more and more will be destroyed. In particular, the tori associated with the periodic, resonant orbits will be destroyed and because of this, the phase point will roam over more and more of phase

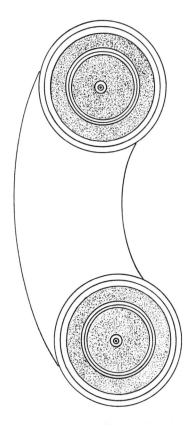

A Poincaré section showing periodic trajectories.

space. It jumps back and forth between unstable periodic orbits, following a given orbit for only a short while before it switches to a different one. Eventually the whole region becomes unstable, and chaos takes over.

If you look at a Poincaré section of the tori after the perturbation has been applied you see a strange complexity. Many of the circles are gone, but you find smaller circles in the region between the larger ones. In fact, you find entire copies of the

A Poincaré section showing what happens to the periodic trajectories. Note smaller version of original section.

cross section. It has become fractal in that the region is now self-similar. The smaller circles contain further copies of themselves, and so on ad infinitum.

The KAM theorem was considered to be an important breakthrough that would lead to a much better understanding of the stability of planetary systems, but when it was applied to the solar system there were difficulties, and it wasn't as helpful as hoped. Nevertheless, it has played an important role in our understanding of long term stability, and was a significant discovery.

COMPUTERS AND CHAOS

What was still needed at this stage was something that would be helpful in performing routine calculations, and it came with the introduction of computers in the 1950s and 1960s. One of the first examinations of long-term stability in the solar system was made in 1965 when several astronomers looked at the stability of the outer planets. Their motion over the next 120,000 years was examined in detail (the study was later extended to a million years). They found no sign of chaos, but they did find that Pluto was occasionally in resonance with Neptune and might eventually develop chaos.

It was 20 years before anyone followed up on this research, but finally, with better computers the study was extended to 5 million years. Still, there was no chaos, but again the long-term stability of Pluto was seen to be questionable.

One of the first positive identifications of chaos in the solar system came, interestingly, not in the study of planetary orbits, but in the study of a moon. It was made by Jack Wisdom, who was then at the University of California at Santa Barbara. In the early 1980s Wisdom began looking into the strange tumbling motion of Hyperion, one of Saturn's moons. It was first noticed during the 1981 flyby of Saturn by Voyager II when the first good closeups of the moons were sent back to Earth. Scientists were surprised in that they expected Hyperion to be tide-locked to Saturn. When a small moon is close to a large planet the gravitational forces across the surface of the moon gradually slow it down until its orbital period is equal to its spin period. This is the case with our moon; it is tide-locked to the Earth.

Once they got a closeup of Hyperion, however, it was easy to see why it wasn't tide-locked. It was oblong, and it is well-known that it is difficult for an oblong object to become tide-locked. The reason is that objects closer to a planet move faster than those farther away. Because of its shape Hyperion would therefore be under considerable stress as part of it (the part closest to the planet) would be trying to orbit faster than parts farther away. Hyperion appeared to be moving erratically and indeed, Wisdom and two colleagues showed that its motion was chaotic. Furthermore they showed that it likely became chaotic quite recently.

As Voyager II continued past Saturn to Neptune, astronomers began focusing on Neptune's largest moon, Triton. Would its motion be chaotic? Peter Goldreich of the California Institute of Technology and several colleagues modeled Triton's history on a computer. According to their results, Triton was at one time in orbit around the sun, but during one of its passes, it came close to Neptune and was captured. Initially it had a highly eccentric orbit which allowed it to cannibalize several of Neptune's other moons. In time, however, perturbations pulled it into a more circular orbit. The only moons escaping the cannibalization were those inside its orbit.

The results from Voyager II seemed to bear out Goldreich's picture. Six moons were discovered, all closer to Neptune than Triton. One of Neptune's other moons, Nereid, is also suspect; it has an eccentric orbit and astronomers believe that it may have passed through a period of chaos in the past. Mars' two moons, Demos and Phobos, may have also passed through a similar period.

Wisdom's interest in Hyperion eventually led to a search for chaos in other parts of the solar system. In the early 1980s he became interested in the asteroid belt. Although most of the asteroids lie between the orbits of Mars and Jupiter, they are not evenly distributed. In the 1850s, Daniel Kirkwood of Indiana University discovered that there are gaps, and these gaps occur where the period of the asteroid is in resonance with the orbital period of Jupiter. One gap of particular interest occurs where the particle's revolutionary period around the sun is in a 3 to 1 resonance with Jupiter's period. There was some indication meteorites that were striking Earth might be from this gap, but there was no proof. Wisdom examined the gap and found Jupiter's perturbation created chaos, which in turn projected particles out of the gap at tremendous speeds. His calculations indicated that some of these objects could, indeed, strike Earth.

Wisdom turned his attention in 1986 to the outer planets. They had been studied earlier, but faster computers were now available and Wisdom realized that it would be possible to push much further into the future. Along with Gerald Sussman of MIT he extended the study to 100 million years (they also looked 100 million years into the past). Again most of the attention centered around Pluto. Its motion seemed so interesting they decided to push further, and over a period of several months went to 845 million years, finding a distinct resonance between Pluto and Neptune, one that could create chaos.

In a similar study of the inner planets, Jacques Laskar of France looked 200 million years into the future and discovered several indications of resonance that could lead to chaos.

A different approach was taken by Martin Duncan of Queen's University and Thomas Quinn and Scott Tremaine of the University of Toronto. Using a computer, they looked at the future of several hundred test particles placed in the space between the outer planets, and found that about half of them became chaotic within 5 billion years.

Interestingly, it's not just the future of the solar system that may be chaotic. Chaos may have also played an important role in the past—in the formation of the solar system. In most of the models of formation it is assumed that the solar system evolved out of a spinning gas cloud. One of the first theories of this type (usually called cosmogonies) was put forward by the French scientist Pierre Simon Laplace. According to Laplace, a spinning sphere of matter flattened into a uniformly rotating disk which eventually broke up into rings; these rings condensed to form the planets.

James Clerk Maxwell looked into Laplace's model and showed that it wouldn't work. According to his calculations, the largest object that would be created would be asteroid-sized. It was a catastrophic blow for evolutionary models, and they soon fell out of favor. In the mid-1940s, however, astronomers returned to them, when C.F. von Weizacker found he could get around Maxwell's criticism by postulating turbulence in the gas cloud. He set up a model using a series of regularly spaced vortices (regions of whirling, turbulent gas).

Gerard Kuiper went a step further in 1951; he assumed the distribution of vortices and eddies was random. In addition, he introduced the idea of accretion (accumulation of matter) into his model, and showed that planetary-sized objects could arise as a result of this accretion. A variation of his model is the one that is accepted today. What is particularly important about this model is that it is based on vortices and eddy currents, which are widely known to be associated with chaos.

Astronomers are still a long way from formulating an entire model of the creation of the solar system based on chaos. There is still, in fact, considerable controversy over the importance of chaos in the formulation of the solar system. It seems likely, however, that it played some role. For example, it may have been important in clearing out certain areas in the spinning gas cloud, just as certain areas are still being cleaned out in the asteroid belt, but beyond that little is known for sure.

So far we've only talked about chaos in relation to planets, moons, and asteroids, but as it turns out chaos isn't restricted to these objects. It has recently been shown to arise in the particles from the sun that are trapped in the Earth's magnetic field. These particles, mainly electrons and protons, make up the solar wind. As this wind sweeps by the Earth it is influenced by the Earth's magnetic field, and some of it goes into what is called the magnetosphere. In the direction away from the sun a long magnetotail develops, and the particles become trapped in it. When the sun is particularly active, as it is every eleven years, this magnetotail becomes distorted; it lengthens and narrows. The Earth, however, pulls it back in place, and as a result the particles are thrown toward Earth creating the Aurora Borealis (northern and southern lights).

Sandra Chapman and Nick Watkins of the University of Sussex have looked into the dynamics of this phenomenon, and have calculated the paths of particles as the magnetic field changes. They showed that the particles would spiral along the magnetic field lines, moving as if they were in a coil of wire. At the same time, however, they would bounce back and forth across the narrow magnetotail. Each of these motions occurs with a particular frequency, but as the magnetotail is stretched, Chapman and Watkins showed that the bounce frequency approaches the spiraling frequency and resonance sets in. The paths of the particles quickly become unstable and unpredictable as chaos sets in.

Another place where chaos may play an important role is in the sun. It is well-known that the sun goes through an eleven year activity cycle where the number of sunspots changes significantly. Over a longer period of time there are also other cycles associated with the sun's activity. Looking at a plot of sunspots you can easily see a cycle of approximately 80 years. There are also two well-known sunspot minima; one is called the Maunder minimum and the other the Sporer minimum. There were literally no sunspots during these eras.

A team of astronomers consisting of Michael Mundt, W. Brian Maguire, and Robert Chase of the Colorado Center for Astrophysical Research studied the sunspot cycles from 1749 to 1990, looking for signs of chaos. They made an important discovery. It had always been assumed that a theory of the activity cycle of the sun would be so complex that it would require a large number of equations to model it properly. The Colorado team showed, however, that the cycle could be modeled using three relatively simple equations from chaos theory, and chaos therefore plays an important role in the cycle. Their calculations showed that these equations predicted most of the things that are observed, something no other theory can do. This new theory may give us interesting new insights into the internal makeup of the sun.

Chaos is also likely important in the rings of Saturn. They are, after all, similar in many ways to the asteroid belt, and if there are resonances within the rings, chaos could easily develop. The great red spot of Jupiter may also be chaotic.

The solar system obviously has considerable chaos throughout it. In this chapter we have seen some of the areas it touches. In the next few chapters we will look into it in more detail.

9

Chaos in the Asteroid Belt

*I*n 1772, *Johann Titius, a German astronomer, came to Johann Bode at the Berlin Observatory with a simple mathematical relationship* he had discovered that appeared to give the orbits of the planets. He pointed out that if you start with the series 0, 3, 6, 12, 24, . . . , add four to each, then divide by 10 you get the distance to each of the planets in astronomical units (the distance between the Sun and Earth). Bode was intrigued with the relationship, and spent so much time popularizing it that people eventually began thinking he had devised it and named it after him. Titius was completely forgotten.

But the law (it is, of course, not really a law) was flawed. It predicted a planet between Mars and Jupiter, and no one had ever found one in this region. Several astronomers, however, had suspected there might be one, and when Bode's law became known the region began to attract considerable attention.

Nothing was found, however, until 1801 when the Italian astronomer Giuseppe Piazzi discovered an object he later called Ceres, which was shown by Johann Gauss to be between Mars and Jupiter. The lost planet, it seemed, had been found at last. But within the next few years three more similar objects were found in the region, and all four were extremely small compared to the other planets. The largest, Ceres, was estimated at the time to be only a few hundred miles in diameter (we now know it has a diameter of about 485 miles).

After the first four, almost 40 years passed before others were found. But as telescopes and star maps improved, astronomers began to find more, and by 1850 thirteen were known. What were they? There was only supposed to be one object in this region—a planet. Could they be the remnants of a shattered planet? No one was sure, but many were caught up in the excitement and asteroid hunting soon became fashionable. Anyone discovering an asteroid and determining its orbit to make sure it had not been discovered earlier, had the honor of naming it.

Daniel Kirkwood was one of those caught up in the excitement. Born on a farm in Maryland in 1814, Kirkwood had little formal elementary education, but was enthusiastic about learning. Over a period of years he taught himself mathematics and science. His special interest was mathematics, but he also had a strong interest in astronomy and decided that a good compromise was to apply his knowledge of mathematics to astronomy. He eventually got a job teaching mathematics and science at Delaware College. He later taught at Indiana University and Stanford.

Asteroids first caught his attention in the 1850s. Over 50 were known by then and he soon became intrigued with them. Unlike most others he was not interested in discovering new asteroids; he was interested in their distribution in space. Were they uniformly distributed? He decided to find out. Most had elliptical orbits, with major axes (maximum distance across the ellipse) pointing in various directions, so he had to be careful in plotting them. A random plot would not be meaningful. He therefore plotted the semimajor axis (one-half of the major axis) of the 50 or so that were known, discovering to his surprise and delight that there were several gaps in the distribution.

By 1866 the known number of asteroids had jumped to 87. Kirkwood added them to his plot, but the gaps remained. Furthermore, by now he had made an important discovery about the gaps; they were not random, as first appeared, but occurred where the orbital period of the asteroid was in resonance with Jupiter's period. A particularly large gap occurred at a position where the asteroid's period of revolution around the sun was three times that of Jupiter. We refer to this as the 3:1 resonance. Gaps also occurred at the 2:1 resonance, the 5:2 resonance, and others.

Kirkwood was sure the gaps were related to Jupiter's gravitational pull on the asteroids. It was easy to see that the orbits could be unstable. An asteroid that lines up with the Sun and Jupiter when it is at its farthest point from the Sun (aphelion), for example, is not in a stable orbit. The two objects pass relatively close to one another here, and the gravitational pull on the asteroid from Jupiter is greater than at any other point in the orbit. In a 2:1 resonance, for example, the asteroid, which would have an orbital period of 6 years, would pass relatively close to Jupiter every 12 years (the period of Jupiter) and would be affected by its strong gravitational field.

Kirkwood was convinced that Jupiter was responsible for the gaps, but he couldn't prove it. He had no mechanism for removing them, and as a result the gaps remained a mystery. For over a century astronomers worked to explain them, but to no avail.

The number of known asteroids continued to grow, and when photography was introduced into astronomy in 1891 astronomers began discovering large numbers of them. They move

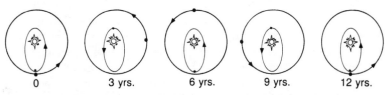

A 2:1 resonance. The asteroid has a period of 6 years and is close to Jupiter once every 12 years. Jupiter is in the outer orbit.

relative to the stars and therefore appear as streaks on a photographic plate. Many of the streaks were, of course, due to known asteroids. An astronomer could only be sure it was a new asteroid if he or she followed it for several nights and determined its orbit. This involved a considerable amount of work, but anyone willing to do it had the honor of naming the asteroid. At the present time several thousand have been named, and a catalog of them has been published. (An interesting new astronomy game has come about because of this catalog. William Hartmann refers to it in his textbook on astronomy. The object of the game is to make up a sentence using only asteroid names. One of Hartmann's favorites, he says, is "Rockefellia Neva Edda McDonalda Hambergera.")

In time most astronomers tired of chasing every streak they saw on photographic plates. Considerable time was required to follow it over several nights and plot its orbit. Furthermore, in many cases the work would be in vain; it would turn out to be a known asteroid.

No one is sure how many asteroids there are, but it may number in the millions. Of those, about 250 have diameters over 65 miles, and a few thousand are bigger then a mile in diameter. Most of the asteroids are between Mars and Jupiter, but a few come within the orbit of Mars, and a few even pass Earth. The ones that come within Earth's orbit are referred to as the Apollos. With so many it might seem that the region is full of asteroids, but this is far from the truth. They are separated, on the average, by more than a million miles. Several satellites have passed through the region without damage.

The questions that were on everybody's mind after the first asteroids were discovered, and for many years thereafter, were: Where do they come from? What caused them? Many thought they had to be the remnants of a shattered planet. It seemed to be a reasonable hypothesis in light of Bode's law, but as astronomers learned more about the solar system, particularly its origin, the idea fell out of favor. Although there are perhaps as many as a million of these objects in the asteroid belt, their total mass is less than 1/50th the mass of Earth, so at best they would make a planet considerably smaller than any in the solar system. Furthermore, as you move outward across the asteroid belt, you find it varies in composition; the asteroids close to Mars are generally much lighter in color than those farther out. The outermost ones, in fact, are almost the color of coal, and likely have high carbon content. This seems to indicate that they were formed when the solar system formed (5 billion years ago). It's well-known that the distribution of elements varies as you moved outward in the solar system, and this same variation appears to be present in the asteroid belt.

Many people tried to explain the gaps after Kirkwood's initial attempt. Some continued to explore Jupiter's gravitational effect on the asteroids, believing it was the key. Others were convinced that collisions had to be involved; if perturbations occurred, the elongation (eccentricity) of some of the asteroid orbits would be changed, and they could easily collide with other asteroids. Still others believed that they needed no explanation; they were convinced the gaps were formed naturally as the solar system itself formed.

Eventually interest began to center on the stability of the asteroids against chaos. As we saw earlier several studies of chaos within the planets had already been made. Was it possible that chaos played an important role in creating the gaps? An answer to this could only be obtained through extensive computation, and few were willing to put in the necessary work. Newton's equations had to be solved again and again. Poincaré made one of the first studies. He looked into the dynamics of the 2:1 resonance. This is the region where the asteroids orbit the Sun twice for each of Jupiter's orbits. Like most early researchers, he applied what is called the averaging method. In this method deviations over short periods of time average to zero and are of little consequence; only long-period trends show. The averaging principle seemed logical, and there were strong arguments that it was applicable, but Poincaré eventually became suspicious of it. He worried about the large changes that could occur with small changes in initial conditions, something we now know is associated with chaos.

It wasn't until the 1970s that astronomers were able to look seriously into the problem of chaos in the asteroid belt, and what allowed them to do this was the development of high speed computers. R. Giffen of Germany made one of the earliest studies in 1973 when, as Poincaré had done earlier, he looked into the stability of the orbits around the 2:1 resonance. Giffen found chaos and suggested it might be associated with the gap, but he presented no mechanism for producing it. More detailed studies were made by C. Froeschle and H. Scholl of Nice, France in 1976 and 1981; they followed the orbits in the region of the 2:1 resonance for 20,000 years, and saw little change. They found chaos in the region, as Giffen did, but concluded that it was not sufficient to cause the observed gap.

Aside from the 2:1 resonance, one of the most conspicuous gaps is that at the 3:1 resonance. In this resonance the asteroid orbits the sun roughly every four years (compared to 12 for Jupiter), and it is 2 1/2 times further from the Sun than Earth.

You can see from the diagram that the asteroid and Jupiter are in opposition every 12 years, creating an unstable situation. It isn't as bad as the one in the 2:1 resonance since it occurs when the asteroid is near perihelion (distance of closest approach to the Sun), nevertheless it was possible that it could lead to chaos.

A graduate student at Caltech, Jack Wisdom, became interested in the 3:1 gap in the late 1970s. While looking into the

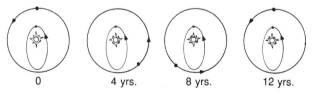

A 3:1 resonance. The asteroid orbits the sun every 4 years. It approaches Jupiter closely once every 12 years.

problem of how it was created, he became convinced that Froeschle and Scholl had not studied the orbits in the vicinity of the gap over a long enough period. Astronomically speaking, 20,000 years is short. In fact, it is only roughly double one of the fundamental periods associated with the orbits. Wisdom knew, however, that with the computers available he was limited. What he needed was a new, faster technique for following the orbits, and he found one in a paper that had been published by the Russian B. V. Chirikov in 1979. Chirikov had used the technique in his analysis of the transition to chaos of charged particles in fusion plasmas. In this method the trajectories are plotted in phase space and Poincaré sections are taken at some point along the trajectories.

Wisdom tailored and polished the technique for use with the asteroids, and to his delight he found that it ran a thousand times faster than traditional programs (programs where Newton's equations were integrated). But was the method accurate? How good a representation of the true asteroid orbits would it give? This was a serious worry and Wisdom spent a lot of time thinking about it.

Astronomers are generally uncomfortable with orbital calculations that are not reversible; in other words, after projecting an asteroid's orbit into the future, you should be able to reverse the procedure and come back to the starting point in the original orbit. Wisdom knew his method was not reversible, except possibly over relatively short periods of a few thousand years. The major problems were the rounding off of numbers in the computer and the rapid divergence of initial neighboring trajectories which results from chaos.

Wisdom therefore set out to verify the validity of his method. He solved the differential equations in the usual way over periods of a few thousand years and compared the results with his much faster method. The agreement was good, convincing him that his method was valid. He then populated the region near the 3:1 gap with 300 fictitious, massless asteroids and calculated their orbits. His aim was to examine these orbits far into the future to see if they remained stable. Each of the asteroids was started at a different point, and therefore had slightly different initial conditions. How far into the future would he have to take them to see if they were chaotic or not? That wasn't an easy question to answer, but as a minimum he would have to cover a period longer than any of the fundamental periods associated with the motions of the asteroids. The longest of these was the precession of its elliptical orbit, the time for its major axis to move through three hundred and sixty degrees. This was 10,000 years, and it seemed that several times this would be needed as a minimum. Wisdom soon found, however, that his new technique was capable of much more than this.

For hundreds of centuries the orbits in Wisdom's study appeared to remain the same, but his computer ground on to 100,000, then 200,000 years. Finally, after about a million years things began to happen. The shapes, or eccentricities, of some of the orbits began to change; they became more elongated. The only ones that changed significantly, however, were those in regions of the phase space that were chaotic. Nearby orbits in stable regions generally remained unchanged. These sudden changes were usually short-lived, but a few hundred thousand years later they would occur again. They appeared as spikes in Wisdom's graphs.

It seemed strange that an asteroid could orbit for a million years in an apparently stable, nearly circular orbit, then suddenly change to an eccentric orbit that took it out past the orbit of Mars. Wisdom was, in fact, so surprised by the result he found it hard to believe at first. His initial suspicion was that the changes were not real, but were caused by the mathematical technique he was using. He checked everything carefully, however, and convinced himself that this wasn't the case. He was even able to verify some of the changes using the slower integration technique.

The sudden elongations occurred only in regions of chaos, yet earlier Froeschle and Scholl had concluded that chaos played little or no role in creating the gaps. Wisdom knew why: their

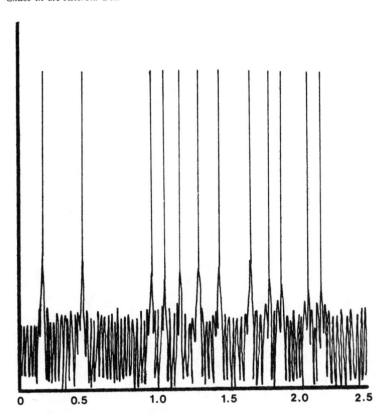

Spikes in a plot of the eccentricity of an asteroid with a chaotic orbit near a 3:1 resonance with Jupiter.

study, he was sure, covered too short a period of time. Looking at his plot, which covered two million years, he could see how they could be fooled into dismissing chaos as unimportant. "Note what a poor idea of the trajectory might be gained from a 10,000-year integration," he wrote in *Icarus*. "Between about 30,000 and 50,000 years the trajectory even stays on one side of the resonance. If that behavior was observed in a numerical survey the orbit would probably have been classified as non-resonant. . . . Viewed over 300,000 years the motion is quite different."

Once Wisdom realized that large changes in eccentricity could occur, it was easy to see how the gaps could be created. The new orbits of the asteroids would carry them past the orbit of Mars and collisions or even near collisions with Mars would eventually occur. The asteroids would be knocked into new and entirely different orbits. The 3:1 gap was therefore not a direct consequence of the gravitational pull of Jupiter, but an indirect one, with Mars also playing a role. Once the asteroids began repeatedly crossing Mars' orbit it was only a matter of time until they were deflected from their orbit. In fact, all the asteroids in the vicinity of the 3:1 resonance could be deflected into new orbits, leaving a gap similar to the one observed.

How long would it take to create such a gap? Wisdom showed that it could easily be done within the age of the solar system, namely 5 billion years. When the calculation was later made more realistic by including three-dimensional motions and perturbations from the other planets, the eccentricities became even larger. Some of the asteroids not only passed Mars' orbit, but also Earth's. Wisdom's mechanism for generating the gap now seemed assured.

Interestingly, Wisdom's discovery also pointed the way to an explanation of another outstanding astronomical problem: the source of the Earth's meteorites. Astronomers had suspected for years that meteorites came from the asteroid belt, but they had no proof. Was this the mechanism? Wisdom showed that one in five of the asteroids in the region of the 3:1 resonance would end up passing the Earth's orbit, and this seemed sufficient to supply the observed number. He was sure this was the mechanism, and his confidence was shared by George Wetherill of the Carnegie Institution of Washington who followed up on the details.

Wetherill knew, however, that it was one thing to get an asteroid in a collision course with Earth, and another to prove

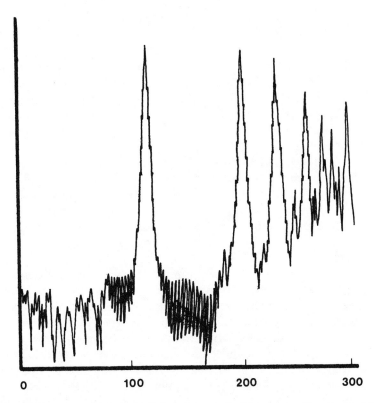

Eccentricity versus time for a chaotic trajectory near the 3:1 resonance. Time (horizontal axis) is in millions of years.

it actually becomes a meteorite. In his effort to prove that this was, indeed, the source of the meteorites, he studied the effects of several processes on the asteroids, effects such as close encounters with planets and fragmentation. To his delight he found that most of the meteorites from this region fell to Earth in the afternoon, in agreement with observation. Furthermore, the numbers he calculated appeared to agree well with observation.

Wetherill then tried several other resonances and found that none of them worked as well as the 3:1 resonance.

But there was a difficulty. The type of asteroid in the region of the 3:1 resonance was one of the most common types known as chondrites, and the composition of chondrite meteorites found on Earth didn't match the composition of the asteroids seen near the 3:1 gap. It is, of course, difficult to determine their composition accurately; we can only study their reflected light. Nevertheless, there seemed to be problems.

Another problem arose when the size of the gap generated by the computer computation was compared to the actual observed gap size. At first the predicted gap seemed too narrow, but Wisdom soon realized that he had compared his data with a sample group in the region of the resonance that was too small. When comparisons were made to a larger group there was good agreement.

We have seen that chaos can cause trajectories of asteroids to suddenly become very elongated. But why does this occur? A better understanding of this can be obtained by looking at the Poincaré section for the asteroids in the region. It is easy to see that it is divided into two parts: clear, nonchaotic regions containing rings of dots and large, chaotic regions covered almost entirely with dots. An asteroid in the nonchaotic region is restricted to a particular orbit, but one in the chaotic region can wander throughout it at random. Since this region is large, orbits can change considerably and you would expect large changes in eccentricity.

Wisdom's success with the 3:1 resonance was considered an important breakthrough. But what about the other resonances? The 2:1 resonance also has a gap; it is much farther from Mars, however, and doesn't appear to be explained in the same way. Even if its eccentricity increased considerably it's unlikely it would cross Mars' orbit. The 3:2 resonance is a more serious problem in that it has an abundance of asteroids, rather than a dearth; they are called the Hildas. The reason for this abundance is still not well understood, but the few calculations that have

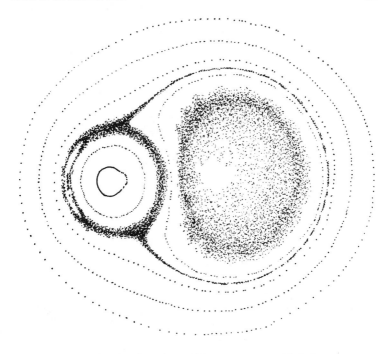

Poincaré section for an asteroid in 3:1 resonance with Jupiter.

been made show little chaos in the region. This may explain why there is not a gap but it doesn't explain the overabundance.

The major reason why the 2:1 and 3:2 resonances haven't been studied in detail is that their dynamics are considerably more complicated than the dynamics of the 3:1 resonance. There are several perturbing influences from objects around them that must be taken into consideration; the 3:1 resonance does not suffer from this problem.

As we saw earlier there is a sizeable chaotic zone near the 2:1 resonance, and it may in some way be responsible for the gap, but we cannot use the same argument we did for the 3:1 resonance.

The region beyond the 3:2 resonance is also difficult to explain. There are few asteroids here. It's almost as if this resonance marks a cutoff, but so far we haven't found the reason for it.

The asteroid belt was a logical place to look for chaos in the solar system, and indeed it was found. Wisdom's explanation of the role of chaos in forging the gaps in the belt was soon universally accepted. But is it the only place where chaos occurs? In the next chapter we will see that it isn't.

10

The Strange Case of Hyperion, and Other Mysteries

*I*n 1848 the eighth moon of Saturn was discovered. Named Hyperion, it was a dim object, so dim that it could be no larger than a few hundred miles across—perhaps 300—making it only about one-tenth the size of Earth's moon. It orbited well outside Saturn's rings at a distance of 980,000 miles from the planet, and its orbit, although eccentric compared to planetary orbits, was considerably less eccentric than many of the asteroid orbits.

Hyperion was just another moon, one among eight whirling around Saturn, and it appeared to have little that warranted special attention. Furthermore, little was learned about it for almost 150 years after it was discovered. Then came Voyager.

In August 1981, as Voyager closed in on Saturn it passed Hyperion, and several photographs of the small moon were taken over a brief period. Astronomers didn't expect to find anything spectacular, and indeed like most other small moons of the solar system it looked like a rugged, heavily cratered rock in space. But it was different in one respect, and this difference was only the first of many surprises that would come over the next few months. It was not spherical as most moons of this size are: It was flattened and looked like giant hamburger 240 miles across by 140 miles thick.

As in the case of other moons, a routine analysis was done using the photographs and other data that were obtained. Of particular interest were its spin period and the direction of its spin axis. Computers were used to compare the orientation of the satellite as seen in the various photographs. The moon had also been studied from farther out, and plots were made of its changing brightness. Coupling this with the computer results, investigators came up with a spin period of 13 days.

But when astronomers tried to determine the orientation of its spin axis, they were amazed. Nothing seemed to make sense.

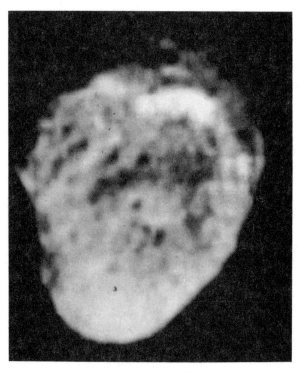

Hyperion. (NASA)

The object was irregular, and they therefore expected it to be spinning around either its shortest or longest axis (they are the only two stable axes). But it wasn't spinning around either of them.

Also surprising was that Hyperion's spin period (13 days) was quite different from its orbital period (21 days). In most moons these periods are the same; we refer to it as synchronous motion. Our moon is a case in point; it is synchronous and because of this we see the same face night after night. Before we ventured into space, we had no idea what the other side of the moon looked like. [We now know that it is similar to the near side but has virtually no seas (dark areas).]

It's relatively easy to see why synchronous motion occurs. Consider our moon. One side of it is considerably closer than the other (2160 miles closer, to be exact) and since gravity falls off with distance, the nearest side is being pulled with a greater force than the far side. This creates what is called a tidal bulge on the Earth in the direction of the moon. (There is also a similar bulge on the opposite side.) The oceans of the Earth, and even the land areas, are pulled outward.

As the moon moves around the Earth this bulge tends to follow it. But, strangely, it isn't able to keep up. If you have ever lived near the ocean you may be familiar with one of the consequences of this. High tide does not occur when the moon is directly overhead; it occurs several hours later. What causes this "lag" in the motion of the bulge? It is due to a loss of energy that arises because there is friction between the Earth's oceans and its land masses as the bulge moves.

One of the consequences of this friction is a tiny increase in the length of our day—about 33 seconds a century. And interestingly this, in turn, has another consequence. It is well-known in physics that spin (or more exactly, angular momentum) must be conserved. This means that if there is a change in the spin at one place in a system, it must be compensated for at another place. In the Earth-moon system this compensation takes the form of a slow outward motion of the moon; it is moving away from us by about an inch a year.

Furthermore, just as the frictional force causes a slowing of the spin of the Earth, it also causes a significant slowing of the spin of the moon. In fact, the moon's spin has slowed so much since its birth that it now keeps the same face toward us at all times. Incidentally, people sometimes think that because the moon presents the same face to us, it's not spinning, but this isn't true. You can easily prove this to yourself by holding a book at arms length and moving it around you, keeping the cover toward you. The book has obviously spun on its axis as it moved around you. The same applies to the moon.

Since synchronous motion is common for moons in the solar system, it was expected that Hyperion, which would experience considerable tidal friction from Saturn, would be tide-locked to it. But the Voyager photos showed that it wasn't.

Jack Wisdom, his professor, Stanton Peale, and Francois Mignard of France became interested in the strange antics of Hyperion. At the time most astronomers were concerned with Saturn's largest moon, Titan. It had been known for years that Titan had an atmosphere, and as Voyager closed in on Saturn, astronomers were anxious to see what it was made of. They expected it to be composed mainly of methane and ammonia, and were surprised to find that it was mostly nitrogen. With the intense interest in Titan, tiny Hyperion was almost overlooked.

But when the photographs of Hyperion came in, and the data was analyzed, astronomers were amazed. Hyperion was spinning in a strange way. Nothing like it had ever been seen in the solar system. Wisdom and his colleagues were intrigued with the result. What was causing its strange motion? Setting up a simple computer model of the motion of the satellite, Wisdom and his colleagues found that there were two major reasons for its behavior: Hyperion's odd shape, and the presence of nearby Titan. Hyperion, in fact, was in a 4:3 resonance with Titan—it orbited Saturn three times for four orbits of Titan. Wisdom showed that the large eccentricity (elongation) of Hyperion's orbit was primarily a result of this resonance.

A large eccentricity, it turns out, makes synchronous motion difficult. The speed of a moon in a very elliptical orbit varies considerably as it travels around its orbit. Its orbital velocity is considerably greater when it passes close to its planet, as compared to when it is at its most distant point. Synchronous motion is therefore difficult for Hyperion.

The three men calculated how long it would take Hyperion to come to synchronous motion, and found that it was long compared to most moons; it was roughly equal to the age of the solar system. Their major interest, of course, was to find out why the spin axis was unstable. Was it, in fact, chaotic? They couldn't model the system exactly—Hyperion is too strangely shaped for this—so they approximated it. They began by assuming that its elliptical orbit was unchanging. This was a good approximation because the changes were much slower than the changes in the spin axis. And finally they assumed the spin axis was initially perpendicular to the plane of the orbit. Coupling this with the tidal bulge, its lag, and the resulting frictional force, they set the problem up on the computer and looked at the motion of the moon.

As usual they plotted the trajectory in phase space and looked at Poincaré sections. Stable orbits in a Poincaré section would occur only in clear regions, and chaotic orbits only in regions covered by dots. And what they got surprised them.

Several spin–orbit resonances were evident in the diagram. These are regions where there is an integral relation between the spin period of the moon and its orbital period around the planet. The synchronous motion we discussed earlier has a spin–orbit ratio of 1:1. But many other spin–orbit resonances occur in the solar system. One of the best known is that of Mercury; it is locked in a 3:2 resonance, spinning on its axis three times for every two trips around the sun.

Resonances such as this are usually surrounded by chaotic regions—regions where the spin becomes unpredictable. But in most cases these regions are narrow. A Poincaré map of a moon

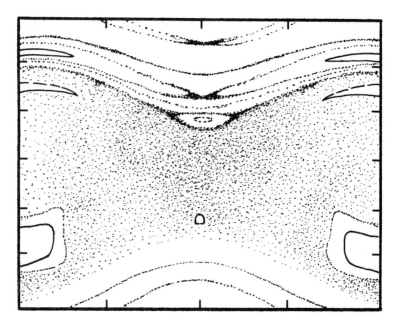

Poincaré section for Hyperion. Islands represent stable motion; dotted areas represent chaos.

may contain several resonances, each surrounded by a narrow chaotic zone, but when the moon is irregular in shape as Hyperion is, the chaotic regions become larger, and eventually begin to overlap. Wisdom and his colleagues showed in the case of Hyperion that the chaotic region around the synchronous state was so large that the 3:2 resonance disappeared and the 1:2 and 2:1 resonances were reduced to small islands. These islands are shown in the figure.

Interestingly, Hyperion could have survived in the synchronous state if its spin axis had remained perpendicular to the orbital plane, even though it is in a large chaotic region. Wisdom showed, however, that once it entered the chaotic state a tiny displacement could start it tumbling erratically.

Looking at the figure we see what is left of the synchronous state; it is in the left-hand corner of the diagram. The island in the upper right is the 2:1 resonance island, and the curves at the bottom correspond to non-resonant quasiperiodic rotation. It is easy to see from the map that unless Hyperion is very close to a resonance, it is going to be in a chaotic orbit.

The most likely scenario for Hyperion's journey to its present state, according to Wisdom, is as follows. At one time it had a rotation period much shorter than its orbital period, and its position in the Poincaré diagram was above the top of the figure. At this time its spin axis was roughly perpendicular to the plane of it orbit, but as time passed it spun slower and slower. If it hadn't been for its odd shape and the presence of Titan, it would have ended in a synchronous orbit. Instead, it became chaotic, and has remained that way ever since.

Chaotic motion is not uncommon in such a situation. As we saw earlier there are narrow chaotic regions around all resonances, including the 1:1 or synchronous resonance, so as resonance is approached the satellite passes through a chaotic region. In most cases, though, the region is narrow and the satellite remains chaotic for only a short time. This, however, did not happen in the case of Hyperion; most of its Poincaré map is chaotic.

Once Hyperion entered its chaotic zone its motion would have been affected dramatically in a short period of time. In as few as two orbital periods it would have started to tumble erratically, and this is what the Voyager images appeared to show. But there were too few images to get a good comparison with Wisdom's predictions. More observations were needed, and they would obviously have to be ground-based.

James Klavetter heard about the problem in 1984 and decided to see if he could get the needed observations. But there was a problem. Successive observations over many night were needed—over as many as 2 to 3 months, and this was a tall order. Time on all large telescopes is at a premium; getting three or four night in succession is an accomplishment; getting 40 or 50 nights in succession is generally unknown.

What was needed was a long-term study of the variation in brightness of Hyperion. This variation comes, of course, from the light that is reflected from its surface as it tumbles; it would easily be detectable in a large telescope.

With a combination of luck and good planning, Klavetter managed to get 37 nights of successive data at the McGraw-Hill Observatory near Flagstaff, Arizona. He had many corrections to make, but when the data was plotted it showed a distinct chaotic pattern in the light variations from the moon, and the variations were consistent with chaotic tumbling.

CHAOS IN OTHER MOONS

With his interest aroused by Hyperion, Wisdom went on to look at other moons. If Hyperion's spin was chaotic it was reasonable to assume that the spin of some of the other moons in the solar system might also be chaotic. Nereid, one of the moons of Neptune, looked like a good candidate; it had a highly eccentric orbit and was similar in size to Hyperion. But a brief investigation showed that it was not chaotic.

Wisdom then turned to Deimos and Phobos, the two moons of Mars. They are small moons; the larger, Phobos, is about 16 miles in diameter, and has an orbital period of 7 hr 39 min. Deimos, which is farther out and smaller, is about 8 miles in diameter and has an orbital period of 30 hr 18 min. Both keep the same face toward Mars; in other words, they are synchronous.

Wisdom plotted Poincaré sections for each of the moons and found chaotic regions in both cases. Neither was chaotic at the present time, but Wisdom believes that both have been chaotic in the past. According to his calculations Deimos likely remained chaotic considerably longer than Phobos, mainly because the time for Deimos to reach a synchronous state is approximately 100 million years, while it is only 10 million years for Phobos.

Another interesting moon is Miranda, one of the moons of Uranus. The Voyager spacecraft passed a mere 24,000 miles from it. Miranda is one of the smaller moons of Uranus, but it is intriguing in that its surface has obviously been modified by internal forces. There are huge valleys—35 miles across and 6 miles deep—produced by some kind of stress across the surface. Large mountains, volcanic cones, lava flows, and huge cliffs are also visible. A number of researchers have suggested the surface features may be due to chaotic tumbling, but detailed analysis has shown that this isn't likely. Stanley Dermott of Cornell University and several colleagues looked into the dynamics of Miranda's orbit in 1988 and showed that it may have been chaotic shortly after its formation, but this was too early to have had an effect on its surface.

CHAOTICALLY SPINNING PLANETS?

If the spin axis of a moon can became chaotic it's natural to ask about the spin axes of the planets. Are any of them chaotic, or more generally, have any of them been chaotic in the past? We know the tilt of the Earth's orbit is 23 1/2 degrees, but has it always been tilted by this amount? And what about the other planets? J. Laskar and P. Robital of the Bureau des Longitude in Paris looked into this in 1993. Using a computer they examined the axial dynamics of each of the planets, and found some surprising results.

Mercury, as we saw earlier, is presently spinning very slowly, trapped in a 3:2 spin–orbit resonance. Laskar and Robital found a large chaotic zone, and showed that the tilt of Mercury's axis may have varied all the way from zero degrees to 100 degrees in the past. According to their results, Mercury's spin rate early on was likely much higher than it is now, but with tidal friction it gradually slowed down. Although the orientation of its spin axis was chaotic, it stabilized as Mercury entered the 3:2 spin–orbit resonance.

Venus is particularly interesting in that it has retrograde motion; in other words, it spins in a direction opposite to that of most of the other planets. Only one other planet, namely Uranus, has motion of this type. Astronomers are still not certain why Venus' spin is opposite that of most of the other planets, but the gravitational pull of the Earth has no doubt been partially, or perhaps totally, responsible. Laskar found a large chaotic region for Venus and showed that the orientation of its spin axis likely underwent large changes in the past.

The most interesting planet in Laskar's survey was Mars in that it was the only planet currently in a chaotic zone. Large variations (zero to 60 degrees) in the direction of its spin axis are possible even now. Laskar showed that it can change drastically in just a few million years. Such changes usually bring considerably more sunlight to the poles than the equatorial regions, and would therefore have to be taken into consideration by anyone trying to understand the climate changes that have occurred on Mars, and their relation to the geology of the planet.

The Earth may also have been chaotic early on, but it was stabilized when the moon was formed, and it is now so stable it is unlikely it will undergo much of a change from its present 23 1/2 degrees in the near future. This is good news since changes as small as 2 degrees can trigger ice ages. Unfortunately, our moon is slowly moving away from us, and the Earth may again enter a chaotic zone in the very distant future.

The outer planets, according to Laskar's study, all appear to have stable spin axes.

THE RINGS OF SATURN

Saturn, with its beautifully complex ring system, is one of the most awe inspiring objects in the sky. Its rings are so much like the asteroid belt we immediately wonder if they also have regions of chaos in them. Like the asteroid belt, they have gaps; two are easily seen from the Earth: the Cassini and Enke gaps. Are they a result of chaos? So far, no one is certain, but it is quite possible chaos played a role in forming them.

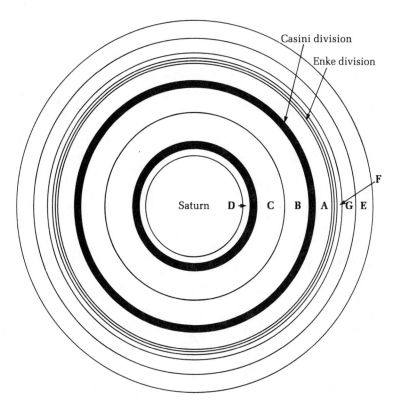

Saturn's rings, showing various regions.

The rings are thin—probably less than a quarter of a mile wide. Composed of ice-covered rocks, they extend out approximately 46,000 miles from the planet. From Earth we see three distinct rings, referred to as A, B, and C. Another ring just outside A was discovered in 1979 by the Pioneer spacecraft; it is called the F ring. Several other minor rings have also recently been discovered by spacecraft.

The brightest of the three major rings is the B ring, which extends out to 21,000 miles. The particles are closely spaced,

Closeup of Saturn's rings showing detailed structure. (NASA)

Closeup of Saturn's rings. (NASA)

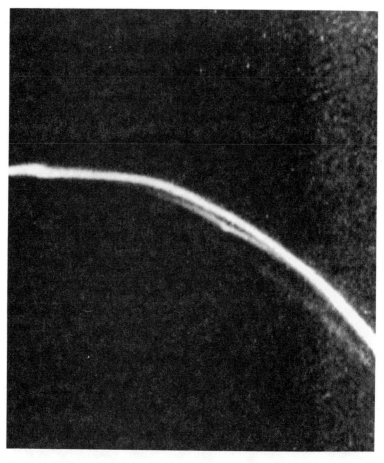

Saturn's narrow F ring showing braids and kinks. (NASA)

making this ring more opaque than the others. Interesting short-lived "spokes" are sometimes seen in it. At the outer edge of the B ring is the 2500-mile wide Cassini gap. This is the gap that is most easily seen from Earth, but surprisingly, when it was photographed by Voyager it was shown to be far from empty. It contained several subrings, and possibly even a small moon.

Beyond the Cassini gap is the A ring. It contains another small gap, called the Enke gap, which is approximately 240 miles across, and is intermediate in brightness between the A and C rings. Beyond A is the isolated, narrow F ring which is 65 miles wide. It has a peculiar braided structure discovered by Voyager. It was the sensation of the Voyager visit, and there was considerable speculation at first that it defied the laws of physics. But two small satellites were discovered, one just outside the belt, and the other just inside. Named Prometheus and Pandora, they are believed to be responsible for the strange structure of the ring; they keep the particles confined, and are therefore sometimes called the Shepherd satellites.

One of the most fascinating discoveries of Voyager was the tremendous substructure of the rings—tens of thousands of ringlets were seen, grouped into dark and light patterns. Distinctly visible in these patterns were spiral density waves. The presence of these waves indicates that the particles of the ring interact with one another gravitationally as they whirl about the planet, each in its own orbit. The ring system is in many ways like a huge rubber sheet, which, in places, has slight wrinkles due to particles being pulled out of it by moons.

There are still many unanswered questions about the rings, and as in the case of the asteroid belt, chaos may play an important role in shaping them. The origin of the Cassini and Enke gaps is still unknown, and since they are similar to the gaps in the asteroid belt, chaos may have played some role in creating them. But there are also other questions. What generates the spiral density waves? Why are some of the ringlets so eccentric? Why is the edge of the ring system so sharp? The answers lie, no doubt, in the interaction between the satellites and the ring particles.

Without the moons the system ring would slowly dissipate; some of the particles would move outward and eventually be lost, others would move inward where they would eventually interact with the atmosphere of the planet. But the ring has obviously lasted for millions of years, and therefore must be relatively stable. And it is the moons that stabilize it. The outer edge of the A ring, for example is stabilized by the moon Mimas. There is a 2:3 resonance between the particles here and Mimas' orbital period.

The Cassini gap may also be caused by Mimas. The particles at its inner edge are in a 2:1 resonance with it; they orbit twice to each of Mimas' orbits. But the details of how Mimas could cause the gap are still unknown. Satellites within the gaps would be of considerable help in clearing them out, and even before Voyager, most astronomers thought there were likely satellites here. A thorough search of both gaps was made by Voyager, however, and no satellites with diameters greater than about 7 miles were found. Nevertheless, there is evidence of a small moon in each of them; analysis of the ring structure shows a "wake" that might have been caused by a moon. And of course it's possible Voyager could have missed a small moon in its search, particularly if it had a dark surface.

Mimas also appears to be responsible for the density waves. It does not orbit Saturn in exactly the same plane as the ring and therefore gravity pulls particles slightly out of the plane. Saturn, of course, pulls them back in, and in the process a wave would be set up.

THE GREAT RED SPOT

Anyone looking at a picture of Jupiter for the first time is amazed by the large orange blemish on its surface. Called the Great Red Spot, it has been visible for over 300 years—literally since telescopes were invented. It has moved slightly over the years, changed color slightly, but for the most part it has remained in the same general region of the planet, and has remained orange-red in color. Before Voyager there were many theories of its origin, most quite speculative. One theory had

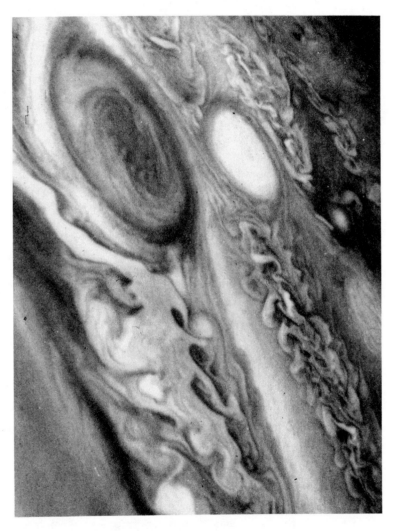

Closeup of the Great Red Spot. Note smaller ovals and disturbance. (NASA)

lava flowing out of a volcano, another that it was the top of a rising column of gas, and a third that it was a gigantic bubble of hydrogen and helium. It appeared to be a giant storm, but it wasn't like any storm we have ever experienced on Earth. First of all it is considerably larger than Earth—8000 miles by 18,000 miles—so it would take two Earths to cover it. Yet it is probably very thin—no more than 30 miles in thickness.

One might think of it as a hurricane of immense dimensions, but hurricanes on Earth rotate counterclockwise above the equator, and clockwise below it. The red spot is rotating in the opposite direction (with a period of approximately 10 hours), so it would be an anti-hurricane.

Many thought the nature of the Red Spot would be resolved once Voyager got closeup photographs of it, and indeed the photographs that were sent back were beautiful. They showed that the center of the spot is relatively quiet, with most of the turbulence at the edge. Many smaller versions of the spot were seen nearby. Using photographs that were taken in sequence, astronomers were able to get spectacular time-lapse movies of their motion. Tiny ovals appeared and disappeared. You could see the Great Red Spot swirling on its axis, like a giant egg in a sea of chaos. But why do the small ovals appear and disappear, while the Great Red Spot remains century after century? The Voyager plates gave us a clue but they didn't give us a definitive answer.

The dynamics of the Great Red Spot became a challenge to those interested in fluid flow since it resembled fluids in motion on a gigantic scale. Furthermore, it could be modeled as a fluid in a computer. One of those who became interested was Phillip Marcus of the Department of Mechanical Engineering at the University of California at Berkeley. After studying the photos from Voyager thoroughly he set up the appropriate equations on the computer. He was particularly interested in seeing if he could produce spots similar to those seen.

The spots that appeared in Marcus' graphs were, indeed, remarkably similar to the spots on Jupiter. Marcus made slides of the spots, then assembled them into a movie. The colors were striking as the vortices changed and merged. Some died and disappeared, others grew. Finally a giant vortex appeared embedded in a chaotic region of the surface. It was the result of several vortices coming together to form a huge one, one so robust and stable it refused to die. It was a region of stable chaos and strongly resembled the Great Red Spot.

Experimental verification of Marcus' analysis came from Joel Sommeria, Steven Meyers, and Harry Swinney of the University of Texas. They simulated fluid flow in a spinning laboratory apparatus and found that large vortices similar to the Red Spot formed in the apparatus. They grew and became extremely stable.

COMETS

We saw earlier that some of the asteroids near Jupiter have become chaotic. Jupiter also influences many comets, and as a result there has been considerable interest recently in the possibility of chaotic comets. B. Chirikov and V. Vecheslavov of the USSR decided to check on Halley's comet in 1987. They developed a simple model of the dynamics and found that its orbit was chaotic as a result of perturbations by Jupiter. Does this mean Halley's comet will leave the solar system? According to Chirikov and Vecheslavov there's little danger of this in the near future; it will continue to visit the sun for another 30 million years.

Another comet in the news lately was Shoemaker–Levy 9, the comet that crashed into Jupiter during the summer of 1994. Calculations also showed that its orbit was chaotic; it was this chaotic behavior, in fact, that caused it to crash into Jupiter.

In this chapter we have seen several more examples of chaos in the solar system. The details have not been fully worked out in some cases, but we are almost certain that chaos is involved. The rings of Saturn, for example, are so similar to the asteroid belt that it seems almost certain that chaos played a role in their formation. At any rate it is a fertile field of exploration. A considerable amount of work has also been done in relation to Jupiter's surface, and it is another region worth exploring further.

So far, however, we have said little about the planets themselves. Are their orbits stable, or more generally, is the solar system stable? We will look into that in the next chapter.

11

Is the Solar System Stable?

*F*or years one of the most formidable problems in astronomy has been the stability of the solar system. What is in store for it in the distant future? Will the planets continue to orbit smoothly as they have for billions of years, or will tiny changes gradually lead to chaos? And if chaos does arise what type of havoc would it cause? Would planets spin off to space? Would the Earth eventually become uninhabitable because of catastrophes, ice ages, and gigantic tidal waves? As we saw earlier many early scientific explorers were concerned with these problems.

A FIRST GLIMPSE OF CHAOS

Pierre-Simon Laplace devoted much of his life to exploring the stability of the planets. With uncanny skill he delved deep into the dynamics of their orbits, his efforts culminating in the publication of five large volumes titled *Treatise on Celestial Mechanics*, a compendium of all that was known about planetary dynamics at the time.

Using perturbation theory, Laplace solved the problem of planetary motion, showing that the planets of a solar system would move quasiperiodically. If plotted in phase space they would trace out trajectories on the surface of a torus. But the solution obtained by Laplace was in the form of a series (e.g., $a + b + c + d + ...$) and for it to be a rigorous, valid, and usable solution, the series had to converge to a finite value. For years no one could prove that it did converge.

Then came Poincaré. He showed that in most cases the series diverged; this would occur, for example when the term b in the above series is larger than a, and c larger than b, and so on. If you sum such a series you get infinity, which is not a useful solution. But Poincaré pressed on, introducing phase space, where solutions are represented as orbits, and he took the bold step of looking at cross-sections in this space, rather than entire orbits. In the process he laid the foundations for a dynamic new approach to orbital theory, an approach that gave him the first glimpse of chaos. He does not appear to have recognized it as chaos, but he knew it was something that rendered his equations too complex for solution.

Few cared to delve into the abyss Poincaré had uncovered. One who did, however, was the American George David Birkhoff. Birkhoff produced rigorous proofs of some of Poincaré's conjectures, and he looked carefully at the startling discovery Poincaré had made. Furthermore, he was one of the first to examine the properties of attractors and attracting sets. Vladimir Arnold of the USSR also made an important contribution to the problem, showing that an ideal planetary system consisting of small planets would be stable. Our system, unfortunately, didn't satisfy his conditions.

In all early models, analytic methods (using algebraic expressions) were used, but with the arrival of computers, numerical methods began to take over. Analytic equations were still required, but the brunt of the work was done by the computer. Computers were a magic wand, allowing investigators to probe millions of years into the future.

But what would we expect to find? It's obvious that the solar system is relatively stable; we see little evidence of catastrophes other than those associated with its formation (many of which have left their scar on the moon). Any changes that occur must be small and subtle, and take place over a long period of time, so we have to be able to project millions and even billions of years into the future (and past) if we are to see them.

Although early computers freed us from hours of tedious calculations, and were extremely useful in astronomy, they were not fast enough to allow a detailed study of the entire solar system. Because of this, investigators restricted themselves to parts of the solar system; in the first studies only the outer planets were studied.

DIGITAL ORRERY

Prior to 1983 astronomers had projected about five million years into the future. That was the year Gerald Sussman, an electrical engineer at MIT, took a sabbatical and went to Caltech. At MIT Sussman worked on computers, computer logic, and the mathematics of learning and intelligence, but he had a long-standing itch to broaden his frontiers. In 1971 he had taken a course called "Stellar Dynamics and Galactic Structure" from Alar Toomre at MIT, that sparked his interest in astronomy—particularly in the dynamics of stellar and planetary orbits. Toomre was an enthusiastic teacher with a passion for colliding galaxies (he was the first to simulate the collision of two galaxies on a computer), and his enthusiasm spilled over to Sussman.

Sussman's major was mathematics, but he never forgot Toomre's class, and resolved one day to pursue his interest in astronomy. After graduation Sussman began teaching in the Department of Electrical Engineering and Computer Science at MIT, and when he became eligible for a sabbatical in the early 1980s he thought about astronomy. He went to Toomre and told him about his interest, and asked his advice, "Why don't you talk to my friend Peter Goldreich at Caltech," Toomre said. Sussman did, and a one-year stay at Caltech was arranged for him.

Digital Orrery. (Photo courtesy of G. Sussman)

Goldreich's interest at the time was the dynamics of objects in the solar system. His student Jack Wisdom had just finished his study of the 3:1 resonance in the asteroid belt. Ironically, by the time Sussman got to Caltech, Wisdom had got a job at MIT and had moved there.

Sussman read Wisdom's thesis carefully, and looked over his published paper. He was amazed at the ingenuity that had been employed, but as a computer expert, one aspect of it bothered him: the approximations that had been taken. "I wasn't sure I liked them and I said to myself the only way I'm going to be sure they are reasonable is by doing the integration myself," said Sussman. But there was a problem. The only computers capable of handling such a large problem were supercomputers, and time on supercomputers was expensive—prohibitively expensive for the kind of time Sussman would need. But Sussman had an ace in the hole. For years he had been custom building computers, computers directed at specific tasks. He would design and build a computer that would calculate planetary orbits, but do little more—one with enough speed to solve the type of problem he had in mind.

"I called up several friends at Hewlett-Packard," said Sussman. "They were buddies and I had worked with them on several problems in the past. They mentioned that several new chips that had just been developed might be useful to me." So Sussman, along with several others at Caltech—there were six in all, a mixture of theoretical physicists and computer scientists—built what Sussman eventually called the Digital Orrery. (Orreries are mechanical models of the solar system, named for the fourth Earl of Orrery who constructed the first ones. Some of his models are still on display in the British Museum.)

It was a small device, only about a foot square, but it was powerful—almost a third the speed of a Cray supercomputer, which was the fastest computer at the time. When Sussman finished the device he went back to MIT. Wisdom was now in the Department of Earth Atmosphere and Planetary Sciences and the two men were soon working together. It would be a particularly fruitful partnership; Sussman had a machine that was faster than any other in the world for problems related to the dynamics of the solar system, and Wisdom had several years of experience solving such problems.

Their first project was a check on Wisdom's work on the 3:1 resonance in the asteroid belt, and the Digital Orrery came through with flying colors; it verified everything that Wisdom had done earlier, and it did it in considerably less time. They then decided to look at the outer planets, Jupiter though Pluto. Little had been done on them as a group. They wanted to follow their orbits as far into the future (and past) as possible to see if they became chaotic. The first run took them 100 million years into the future, and 100 million years into the past. It was an exciting journey into time, disappointing in some respects, but extraordinarily satisfying in others. There was no sign of chaos amongst the giant planets, but the orbit of Pluto exhibited some fascinating new features.

Pluto has the most eccentric orbit in the solar system, and it's also inclined to the plane of the other planets, so it was expected that something interesting might be uncovered. Furthermore, it crosses Neptune's orbit. Throughout most of its orbit Pluto is the most distant planet, but at the present time it is actually closer to the sun than Neptune. It might seem that the crossing of the two orbits would seem to open the door to a collision. Strangely, it doesn't. The two planets are in a 3:2 resonance, with Neptune making three orbits around the sun to Pluto's two, and because of this, when Neptune crosses Pluto's orbit, Pluto is far away at the opposite end of its orbit.

The 3:2 resonance was well-known before Wisdom and Sussman's study, but they discovered several other interesting resonances. The inclination, or tilt, of Pluto's orbit oscillates, for example, between 14.6 degrees and 16.9 degrees over a period of 34 million years. More importantly, however, there were indications that Pluto's orbit might be chaotic. The standard technique for checking on chaos is to compare a planet's orbit to one that starts with the same conditions at a slightly different point, in other words, with slightly different initial conditions. The object is to see if the two trajectories diverge, or more specifically, if they diverge fast enough to indicate chaos. In the case of Pluto they did diverge. The planet was so interesting dynamically that they decided to make a much longer run so it could be studied in more detail. But before they could make longer runs several difficulties had to be overcome. One problem was round-off errors. These are errors that arise because of the limit on the number of decimal places that can be carried by a computer. If you make a run far into the future, you should be able to reverse the run and get back to the starting point. Round-off and other types of errors usually prevent this.

Another error that plagued them was one related to the step size. Their run had to be taken in steps; initially they used 40 day steps, which seemed to fit in well with the capacity of the computer, but they found it introduced a large error, so they tried other step sizes. They soon noticed that for certain step sizes the error increased in a positive direction as the run proceeded, and for other sizes it increased in a negative direction. Somewhere in between, it seemed, there had to be a step size that would give almost no error, and with a large number of experiments over almost two years they narrowed in on a step size that gave almost no error—32.7 days.

"It was a weird piece of analysis that no one ever really understood," said Sussman. Furthermore it was peculiar to the Digital Orrery, related to the makeup of the machine, and had nothing to do with the way the problem was set up. But once they discovered it, the route to much longer runs was open. Their previous run of 200 million years, which was the longest ever undertaken, had taken a month. The new run, which again was restricted to the outer planets, took 5 months, but at the end they had projected 845 million years into the future. This was far short of the solar system's age. Nevertheless, it was a tremendous accomplishment.

Looking over the numbers that came out of the computer they again saw little that looked chaotic. Even after all this time the four giant planets still moved smoothly around their orbits; no collisions occurred; the eccentricities didn't change significantly. The run might have been uneventful if it hadn't been for Pluto. Again the major surprises came from its orbit; many new and interesting resonances were discovered. Periodic changes, or variations, that occurred over 3.8 million years, 34 million years, 150 million years, and 600 million years were all clearly visible in the graphs. Still, Pluto at first glance didn't appear to be chaotic; it remained in orbit around the sun, and its eccentricity didn't change significantly. But with a closer look they saw the first evidence of chaos. As they had done earlier, they started a second run with slightly different initial conditions, and watched how fast the two trajectories diverged. A measure of the divergence is given by what is called the Lyapunov exponent. [It is named for the Russian mathematician

A. M. Lyapunov (1859–1918) who was an early investigator of stability in nonlinear systems.] If the motion of the planet is stable and quasiperiodic, the Lyapunov exponent is zero; if a divergence between the two trajectories occurs, the Lyapunov exponent is positive. In this case the distance between the two trajectories doubles, then doubles again in the same time, and so on (more exactly it changes by 2.72; technically we say they diverge exponentially). Thus, the Lyapunov coefficient is a very potent indicator of chaos; if it is positive, chaos is present.

Sussman and Wisdom showed that this doubling in the case of Pluto would occur every 20 million years, which is long by standards here on Earth, but only a tiny fraction of the age of the solar system. It is, in fact, so small that it makes us wonder how Pluto has managed to survive in orbit so long.

How would this chaos effect its orbit? Contrary to what you might think, it doesn't imply that Pluto is going to do strange things. It's never going to collide with Neptune (at least not in the near future), and its eccentricity, while relatively large, hasn't gotten out of control. Pluto has obviously been in orbit since the solar system began, so despite the fact it is in a chaotic zone, it's likely to survive many years into the future without any significant changes.

What chaos implies is not catastrophes, but rather our inability to make long-range predictions about its orbit. An orbit with a large Lyapunov exponent cannot be plotted far into the future; most of our information about Pluto's orbit, for example, is lost in 100 million years.

When I interviewed Sussman I asked him what his reaction was to the discovery of chaos in Pluto's orbit. "The most important thing is always, 'Hey, did we blow it?' and this was our first reaction. We didn't publish any papers without years of analysis. It's so easy to make a mistake. There are all kinds of different errors . . . numerical errors, errors of modeling, errors related to the mathematics, and errors you haven't even thought of."

Sussman has gone on to other problems now—one of his new interests is optical chaos—but he admits there is still lots to be learned about the dynamics of the solar system. As he described and explained his work to me his enthusiasm for the subject was evident. He finished by saying, "My philosophy has always been: You live only 70 years so you may as well have some fun during that time doing the most interesting things you can think of."

Sussman and Wisdom's run was followed in 1989 by one called LONGSTOP. A group at the University of London used a Cray supercomputer to integrate the outer planets for a period of 100 million years. They followed the orbits of the four gas giants, and found that they remained generally stable, with only a slight indication of chaos.

The final computation with the Digital Orrery was made in 1990; new and faster chips were now on the market and its days as a useful computer were over. With some regret and considerable pride Sussman sent it to the Smithsonian Institute National Museum of American History in Washington, D.C., where it is now on display.

Wisdom and M. Holman made another run on the outer planets in 1991 in an effort to look more closely at Pluto's behavior. They followed the evolution of the outer planets for 1.1 billion years, proving again that Pluto's orbit is chaotic. They also verified the long periodic variations associated with the planet that had been discovered earlier.

LASKAR

While Sussman and Wisdom were making their first runs, Jacques Laskar of the Bureau des Longitudes in France was attacking the problem from a different direction. Laskar's initial interest was the orbit of Earth; he wanted to find out what changes had occurred in the past (and what changes might occur in the future), and how they would affect the Earth's weather. Laskar soon decided, however, that it was hopeless to follow the orbits using the usual complicated analytical expressions. He decided to use an averaging process that smoothed out small changes in the orbit, thereby spotlighting only the longer, more significant ones. Despite his attempt at simplification, the expression he used actually contained 150,000 algebraic terms—a monstrosity by any standard.

Because of his averaging technique, Laskar could see only long-term trends, and it made little sense to take short time steps as Wisdom and Sussman had; his increments were therefore a relatively long 500 years. With his program, he was able, using a supercomputer, to project 200 million years into the future, and he did this not only for the outer planets, but all the planets except Pluto.

As Sussman and Wisdom did, Laskar calculated the Lyapunov times for the planets; in other words, he determined how fast two similar trajectories diverged, and found evidence for chaos throughout the solar system. The whole solar system, including Earth, was chaotic according to his computations. This was a surprise. Pluto had been shown to be chaotic, but few expected the entire solar system to be chaotic. Laskar showed that if you started the solar system out with slightly different initial positions, the separation between it and the original system would double every 3.5 million years. This is only a tiny fraction of the age of the solar system (approximately 5 billion years) and means that in as little as 100 million years there would be no resemblance between the two systems. But it doesn't imply that the solar system is in any real danger. Despite the chaos, catastrophic events are unlikely.

Many were critical of Laskar's results when he first published them, and he felt he had to justify them by explaining the source of the chaos. As we saw earlier, resonances are usually the culprit, and in this case they were again to blame according to Laskar. He pinpointed two resonances that he felt were responsible: the first was between Mars and Earth, the second between Mercury, Venus, and Jupiter.

VERIFICATION

Still, Laskar's results had to be verified, and verification, or at least partial verification, came soon. Martin Duncan of Queen's University in Canada, Thomas Quinn of Oxford, and Scott Tremaine of the Canadian Institute for Astrophysics at the University of Toronto had been working for some time on a similar problem using a completely different approach. As part of their program, Duncan and his colleagues ran an integration of the entire solar system using a method quite different from Laskar. It was a more accurate representation of the dynamics and included many more terms, and their run was therefore much shorter. They included corrections for general relativity and for the finite size of the Earth and moon. They computed the long-term changes for the orbits of all the planets. Because of the complexity of their expression their run covered only 6 million years—three million into the future and three million into the past. They were therefore unable to directly verify the chaos found by Laskar, but where the two integrations overlapped there was excellent agreement. In particular, the same resonances were found; one of these resonances was the one that Laskar believed was responsible for the overall chaos of the solar system.

I asked Duncan if he was surprised to find chaos in the region of the outer planets. "At this point Laskar's results were available and people were thinking there would be chaos. So we weren't too surprised," he said.

Duncan and a graduate student, Brett Gladman, also studied chaotic behavior in the outer solar system using direct integrations. Their interest was the identification of regions within the solar system that were not chaotic, regions where undiscovered asteroids and comets might reside. To do this they used 300 test particles, much in the same way Wisdom did earlier in his studies of the asteroid belt. The test particles were massless and therefore didn't perturb the objects around them, but they reacted to the gravitational fields they experienced. Much to their surprise, Duncan and Gladman found that most of the region from Uranus outward in the solar system was chaotic. The orbits of roughly half of the 300 test particles became chaotic enough in five billion years to be ejected from the solar system. Their run wasn't long enough to directly verify Laskar's evidence for chaos, but it showed that many individual orbits in the outer solar system were chaotic, and there was good agreement between the two results over the regions that were common to them.

Duncan got his bachelor's degree from McGill, his M.Sc. from the University of Toronto, then went to the University of Texas at Austin for his Ph.D., where he worked under Craig Wheeler. His thesis was on the dynamics of stars near the center of the giant elliptical galaxy M87, trying to establish whether or not it contained a giant black hole there.

"I've always been interested in the dynamics of orbits. During my Ph.D. I was involved with stellar dynamics, black holes, and galactic dynamics," Duncan said. From the study of stellar dynamics he went to the study of planetary dynamics in the solar system. The physics is, of course, quite similar so it wasn't a large change.

He became interested in the stability of the solar system through an interest in comets. Comets are known to have originated from a shell about one light year from the sun, called the Oort Cloud. More recently a second cloud in the outer region of the solar system—just beyond Neptune—called the Kuiper belt has been identified. Duncan is interested in the dynamics of both of these clouds.

SUPERCOMPUTER TOOLKIT

Further verification of the chaos was still needed, and it came from Sussman and Wisdom in 1992. Digital Orrery had been retired, but Sussman and Wisdom now had a bigger and better computer that they called Toolkit. It was 50 times faster than Digital Orrery, and had also been designed specifically for problems related to the solar system. During the late 1980s computer technology improved significantly and small computers were now capable of doing what only a supercomputer could have done five years earlier. Toolkit therefore had more modern, faster chips in it than Orrery. Furthermore, it was more flexible, yet it wasn't much larger. In reality, it was eight separate computers, each capable of making an independent integration.

With their new computer, Sussman and Wisdom could follow the evolution of all nine planets without serious approximations. Their model was quite similar to that of Duncan, Quinn, and Tremaine, with the exception of their treatment of general relativity. Sussman and Wisdom did not include corrections for general relativity.

Each of the eight computers was started with slightly different initial conditions. They ground on for 100 hours, projecting 100 million years into the future. With this many separate integrations, Sussman and Wisdom could easily see the divergence of the trajectories, and indeed it indicated chaos. Laskar was right, the solar system was chaotic.

Sussman and Wisdom found, however, that the divergences indicated two different Lyapunov times for the solar system: one of four million years and one of 12 million years. The giant outer planets appeared to be dominated by the 12 million years time constant over most of the 100 million years, with the four million year component coming into play only during the last five million years.

What did this mean? The only explanation Sussman and Wisdom could come up with was that there were two separate mechanisms creating chaos in the solar system.

Pluto again appeared to be chaotic with a time scale of 15 to 20 million years, in agreement with their earlier study. Interestingly, the chaos in Pluto's orbit seemed to be independent of the four giant planets.

The verification was a tremendous breakthrough. There was no question now. Three groups, each using a different method, had shown that the solar system was chaotic. Chaos had to play an important role in its evolution. Sussman and Wisdom were, nevertheless, apprehensive. While there appeared to be two mechanisms responsible for the chaos (Laskar had also suggested there were two mechanisms) neither had been positively identified. Resonances between the Earth and Mars, and between Mercury, Venus, and Jupiter had been suggested as a source by Laskar, but Sussman and Wisdom didn't agree. As far as they were concerned the sources had not been positively identified; this was in contrast to the case of the 3:1 resonance in the asteroid belt, and Hyperion, where the cause of the chaos had been pinpointed.

Sussman and Wisdom also worried that the chaos they were seeing was an artifact of their numerical method. But the agreement of the Lyapunov times for Pluto between all three methods—each quite different—argued against this.

But there was an even more serious difficulty: if the entire solar system is chaotic with a time constant of only a few million years, why isn't there more evidence of the chaos? After all, the solar system has been around for about 5 billion years. Is the Lyapunov time really a good indication of the timescale of chaos? Laskar got a Lyapunov time of 5 million years, and Sussman and Wisdom got one of 4 million years, yet the solar system as we know it appears to have been relatively stable for 4.5 billion years. Obviously there must be more to it than the time for exponential divergence. Further study was obviously needed, as well as different methods for measuring chaos.

OTHER PROJECTS

Myron Lecar, Fred Franklin, and Marc Murison of the Harvard-Smithsonian decided to find out why there was little evidence of chaos despite the short Lyapunov times. They studied relatively simple systems consisting of a single test body orbiting the sun and one or two giant planets. In each case they calculated the Lyapunov time and compared it to the time for the test body to cross the planet's path and be knocked out of the system. In each case they found the time for catastrophic events to occur was much longer than the Lyapunov time. Applying their results to the solar system gave a stability time of trillions of years—much longer than the present lifetime of our system.

Duncan and his colleagues at Queens University are working on a problem related to comets that they hope will eventually shed some light on this difficulty. Comets are divided into two types: short period and long period. Comets coming from the Oort Cloud are long period, having periods as long as a million years. Some of these comets, however, are perturbed by Jupiter and Saturn when they are in the inner regions of the solar system; they then take up much shorter orbits entirely within the inner solar system and become short period comets. For many years this was the accepted picture. Then computer studies of the short period comets showed that all of them could not come from the Oort Cloud; some had to come from a region much closer. And indeed as early as 1950 Gerard Kuiper had suggested there was a comet cloud belt just outside Neptune. These studies showed that some of the comets were coming from this belt, now called the Kuiper belt. This has been substantiated recently by observations; more than a dozen objects have now been discovered in it.

Duncan has recently been studying this belt. There is a steady "leakage" of objects from the Kuiper belt, and it therefore has to be chaotic to some degree. We see several objects from this belt passing the sun every year. There are billions of objects in the belt, however, and most of them have been there since the solar system began, so it is, at best, weakly chaotic. "People think there must be resonances related to some of the objects in the Kuiper belt," said Duncan. "We're now capable of following objects in this region for billions of years—almost the entire age of the solar system. We've been doing extensive mappings . . . following objects starting from orbits in the Kuiper belt, looking at what kind of regions are stable, what kinds are unstable, and following the objects as they cross Neptune's orbit, in other words, the influx from beyond Neptune to close to the sun where they become visible. So we're modeling a kind of steady state distribution of comets in the solar system."

Interestingly, Duncan and his colleagues are not using supercomputers for this work. Computers have become so fast in recent years that workstations with relatively small computers are now capable of doing what a Cray supercomputer did only a few years ago. This is a great relief to Duncan; time was expensive on a supercomputer, and difficult to get. With a workstation he can set computers going day and night without worrying about the expense. "We recently got 7 workstations on a grant," he said, "and we've set each of them going for weeks on end. We can follow the four outer giants with hundreds of test particles in a matter of a week or so."

Duncan is looking forward to being able to trace the solar system, or at least part of it, back close to its origin. "You can't actually integrate all the way back to the beginning, but you can use models," he said. He mentioned that Laskar has recently argued that it is possible for Mercury to be ejected from the solar system in 5 to 10 billion years. Duncan said he is interested in following up on it.

There's still a lot of work to be done. Duncan outlined some of the other problems he's interested in: What is the detailed structure of the Kuiper belt? What is the origin of the asteroid? What is their long-term behavior? Could there have been more planets early on? What processes form the planets? Chaos obviously plays some role in all of these, and they are questions people will be looking into in the next few years.

In this chapter we saw that with the discovery of chaos the idealized picture of the solar system as a precise machine has been eroded. We now realize that we live in a system beset with more complexity than we imagined. Chaos has given us a much clearer picture of what is going on.

12

Stars and Galaxies

*C*haos, *as we have seen, is common in the solar system, but recently astronomers have been looking beyond the solar system, out to* the stars, and evidence for chaos has been found even here. Late in its life a star may become unstable and begin to pulse, its surface surging outward, then inward, causing it to brighten, then dim. In most cases the pulsations are regular and periodic, but in some cases they are irregular. In most variable stars, as these stars are called, the changes are controlled, but astronomers have found evidence in recent years for a few in which the changes are unpredictable. In short, they are chaotic. Theoretical models have also shown that stars can pulse chaotically. In fact the same transition to chaos that appears in other systems, namely the frequency doubling we discussed in Chapter 6, appears to occur in certain types of variables.

Systems of stars—galaxies—can also become chaotic. In this case, as in the case of the planets, it is the orbits of the stars that are chaotic. We saw earlier that Hénon found theoretical evidence for chaos in globular clusters; galaxies were therefore a natural candidate, and indeed chaos has been found in several types of galaxies.

CHAOS IN PULSATING STARS

In our search for chaos in the solar system we were dealing with non-dissipative systems—systems that lost no energy—systems in which chaos was present but attractors did not exist. Pulsating stars, on the other hand, are dissipative systems and as we saw in Chapter 5, attractors exist in dissipative systems. The search for attractors is therefore important, and not just for ordinary attractors, but also for strange attractors—those associated with chaos.

To understand how and why stars pulse it is best to begin with their formation. Stars are formed from gas clouds that are composed of hydrogen and helium, with other elements present in tiny amounts. Initially, the gas cloud is huge and irregular, but self-gravity pulls it inward and it gradually becomes spherical. A hazy red sphere forms, but as its core is compressed its temperature climbs until finally it reaches about 15 million degrees and nuclear reactions are triggered. The hydrogen in the core begins to "burn," supplying energy to sustain the star. At this point the inward gravitational fall is balanced by an outward gas pressure and contraction stops. The ball of gas has becomes a star and for millions or billions of years, depending on its mass, it will burn its fuel peacefully with few external changes.

But the "ash" from the burning hydrogen, namely, the helium, is heavier than hydrogen and it accumulates at the center, and like hydrogen it is compressed and heated. When its temperature reaches approximately 100 million degrees it too is ignited. In an average-sized star like our sun it is ignited explosively and the core of the star is blown apart. The hydrogen burning, which is now taking place in a shell around the helium core, is extinguished, and the star dims. Gradually, however, the helium falls back to the center and begins burning peacefully as does the hydrogen in its ring around the helium.

Stars remain in equilibrium throughout most of their life. The inward pull of gravity is balanced by an outward force due to the pressure of the hot gas. Late in the life of the star, however, this equilibrium can be upset, and the star may start to pulse. This doesn't happen in all stars, only in those slightly more massive than our sun. The star expands and contracts radially, usually regularly, but in some cases irregularly, and as it expands its brightness increases for a few days, then the expansion stops and it begins to shrink and grow dimmer. It goes through this cycle again and again. The star has become a variable.

One of the easiest ways to understand their pulsation is to think of a star as a cylinder and piston. Assume the piston is

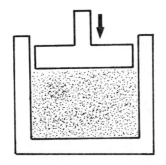

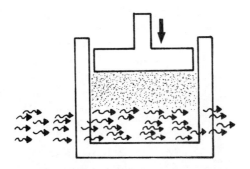

A simple analogy to a star: a piston compressing gas. Lower diagram shows radiation entering and leaving the system.

massive and exerts a downward force due to gravity. It therefore compresses the gas in the cylinder, but the gas exerts pressure and can be compressed only so far. When the downward force exerted by the gravity on the piston equals the upward force due to the pressure of the gas, the piston comes to rest in what we refer to as its equilibrium position.

You can, of course, compress the gas further by pushing on the piston. If you do this and let it go it will oscillate around the equilibrium position. If there were no damping the piston would continue to oscillate forever, but restoring forces such as friction cause the oscillations to damp out, and the piston finally comes to rest at its equilibrium position.

You have a similar situation in a star. If the equilibrium is disturbed, say, by compression, it will begin to pulsate with its surface moving in and out around its equilibrium position. Calculations show that these pulsations will damp out in about 80 to 100 years, depending on the mass of the star. But most of the variables in the sky are known to have been pulsating much longer than this, so there must be more to the pulsation mechanism than a simple loss of equilibrium. There has to be something that keeps the pulsations going.

One of the first to look into this was Arthur Eddington of England; he discussed it in his classic book on stellar evolution. To understand his explanation it is best to go back to the piston and cylinder. Assume now that the piston is transparent to radiation so that the gas can absorb radiation (energy) from the outside.

As the piston moves downward from its equilibrium position the gas is compressed and its pressure and temperature increase. The gas is now more dense, and most gases are able to absorb more radiation as the density increases. Assuming this is the case the gas in our cylinder will absorb more radiation than it did when it was in equilibrium. The radiation will heat the gas and it will begin to expand. As it does, it will push the piston upward. Because of the momentum it gains it will not stop at its equilibrium position, but will continue on upward until the density of the gas is low. When this happens the gas becomes transparent and lets more radiation pass through; in other words, it absorbs little. The piston therefore stops and begins to move downward.

We have a similar situation in a star. In the case of the star the source of the radiation is the nuclear furnace at its center. When the outer shell of the star is compressed this shell can absorb more radiation from the center than it would normally, letting little pass through. As in the case of the gas in the piston, this causes the star to expand until it finally becomes transparent to radiation. The radiation then passes through the shell without being absorbed, and the gas loses energy and begin to contract. This temporary, cyclic "damming up" of the radiation causes the star to increase and decrease periodically in size, and therefore in brightness; in short it becomes a variable.

In most variables the pulsation rate is regular. For hundreds and even thousands of years the star brightens and dims regularly, but as we will see, under certain circumstances this rhythm can be destroyed and the pulsations can become chaotic.

Variable stars have been known and studied for hundreds of years. One of the first was noticed by the amateur astronomer John Goodricke, the son of an English diplomat serving in the Netherlands. Despite deafness and a short life—he lived only to 21—Goodricke made two important discoveries. In 1782, when he was only 18 years old, he began studying the star Algol. He noted that its brightness varied, and plotted its light curve, but he did more than merely observe it; he gave an explanation of its light variation. He explained that it was a binary system, with one of the stars eclipsing the other. Years later it was shown that this was indeed was the case; the star is what is now called an extrinsic variable.

Goodricke was also the first to study another type of variable. In the constellation Cygnus he found a star that showed a different type of light variation—much less abrupt; we refer to it as δ Cephei. Goodricke plotted its light curve, found that it varied with a regular five-day cycle, and was two and a half

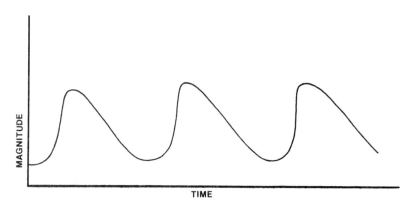

Light curve of a Cepheid variable.

times brighter at maximum than at minimum. Later it was discovered that the light variation of δ Cephei was not due to an external phenomenon (e.g., an eclipse) as in the case of Algol, but due to the pulsations we discussed earlier; it is now referred to as an intrinsic variable.

δ Cephei was the first of a class we now refer to as Cepheids. Cepheids have a period (time to go through their light cycle) between one and approximately 50 days; they are easily distinguished by the form of their light curve—they brighten more rapidly than they dim.

Cepheids are important because they can be used as distance indicators. The distance to a star cluster or galaxy, for example, can be determined if a Cepheid can be found within it. All that is needed is the Cepheid's average brightness and the period associated with its light variation. Because of this, Cepheids have played an important role in the history of astronomy. The relationship between a Cepheid's period and brightness (luminosity), which is referred to as the period–luminosity relation, was discovered by Henrietta Leavitt in 1912. The daughter of a minister, Leavitt graduated from what is now Radcliffe College in 1892. When she graduated she was hired by the director of Harvard Observatory, Edward Pickering, who gave her a job examining plates taken at a Harvard outpost station in Peru. Her main task was to identify variable stars, which she did by comparing plates taken several days apart. It was tedious work and she was paid meager wages. In fact, for the first while she worked for nothing.

Some of the plates were of a pair of irregular southern galaxies called the Magellanic Clouds. Because they are relatively close—they are our nearest intergalactic neighbors—we can see individual stars in them (in most galaxies this is not the case). Leavitt began finding Cepheids in the Magellanic Clouds and soon noticed that the brightest ones had the longest periods, and since all the stars in them are, to a good approximation, the same distance from us, it meant that intrinsically brighter Cepheids had longer periods. Leavitt published her result in 1912. But there was a problem. For the discovery to be useful, astronomers had to calibrate the scale, and to do this they had to determine the distance to a Cepheid independently.

Harlow Shapley of Mt. Wilson Observatory was one of the first to see the importance of Leavitt's discovery. He was interested in determining the size and structure of our galaxy, and our position in it. Using a crude statistical procedure he was able to determine the distance to a number of Cepheids in our galaxy, and using them he calibrated the period-luminosity relation. Using it he showed that the sun was not at the center of our galaxy, as had been assumed for years; it was actually out in the arms, about three-fifths the way out from the center. Later Edwin Hubble used the relation to determine the nature of, and distance to, a number of nearby galaxies.

Another type of variable is called the RR Lyra. As with the Cepheids they are named for the brightest of the group, which is in the constellation Lyra. RR Lyraes pulse with periods less than one day and their variation in brightness is considerably less than that of Cepheids—on the average only about a tenth. They don't exhibit a period-luminosity relation as Cepheids do, but they all have approximately the same intrinsic brightness, and therefore, like Cepheids, are useful as measuring rods. To see why, consider a field covered with 100 watt light bulbs, and assume we know the distance to the nearest one. Knowing the distance to one it is easy to determine the distance to the others since the dropoff in light intensity is the same for all of them. This also applies to the RR Lyraes.

Another class of variable is the large red variables. The best known of this class is Mira, or Mira the Wonderful as it was called by the ancients. Large red variables are considerably larger than Cepheids and RR Lyraes, and have much longer periods. They also change considerably more in brightness. Mira, for example, disappears from view for several months, then becomes almost as bright as the stars in the big dipper. The periods of large red variables vary between 50 and 700 days.

These are the three main types of variables, but there are others, two of which are of particular importance in relation to chaos. One is a subclass of Cepheids, sometimes called type II Cepheids, but usually referred to as W Virginis Stars. They differ from ordinary Cepheids in that they are generally older stars, and have a slightly different shaped light curve. The dimmer ones have a single period, but the brighter ones exhibit multifrequency periods (two or more periods in the same light curve), which is of interest in relation to chaos.

Finally, we have the RV Tauri stars. They are also related to Cepheids, but differ from the usual Cepheid in that they don't have regular periods. They are semiregular variables.

The pulsations of variable stars are similar to the oscillation of a rigid rod, or the swing of a pendulum, and as we saw earlier there is a technique for searching for attractors in one-dimensional plots of amplitude versus time. We discussed the technique in relation to the drops from a faucet used by Takens and his group at Santa Cruz.

Using a plot of the light variation of a star you can set up the phase space and look for evidence of a strange attractor. Furthermore, you can take Poincaré sections as we did in an earlier chapter. The technique is basically the same regardless of whether the object is a vibrating rod, a dripping tap, or a pulsating star.

One of the first to look for chaos in pulsating stars was Robert Buchler of the University of Florida. Born in Luxemburg, Buchler came to the United States for graduate work where he worked on the many-body problem under Keith Brueckner at the University of California. In his thesis he applied the techniques of many body mechanics to dense matter. Upon graduation he took a postdoctoral position at Caltech where he worked for No-bel Laureate William Fowler. His interest in astronomy was sparked here when he began applying his knowledge of dense matter and the many-body problem to neutron stars.

"I was curious about chaos and began reading about it," he said. "I was sure it could be applied to stars." Most of his work on chaos has been theoretical—working with computer models of pulsating stars and seeing under what conditions they become chaotic—but he has also spent time searching for evidence of chaos in the available data.

Just as Feigenbaum varied a parameter and showed that pe-riod doubling occurred en route to chaos, Buchler has shown that a similar phenomenon occurs in stars. "When you change a parameter such as surface temperature and look at the pulsa-tion behavior you see it change gradually. It goes from period two to period four and so on to chaos," he said.

Buchler has been particularly interested in W Virginis stars and RV Tauri variables. In a paper he published with Geza Kovacs in 1987, he showed that computer models of these stars exhibited a series of period doubling bifurcations in their tran-sition to chaos as their surface temperature was varied. He stated that the probable cause of the chaos was a resonance in the pulsations, similar to the resonance we saw earlier in the solar system.

I asked Buchler how well the results of his computer model agreed with observations. He hesitated, then replied, "There are some data, but suitable data sets are hard to find. During the last six months we have analyzed a number of RV Tauri stars using data we obtained from the American Association of Variable Star Observers and we found good evidence for chaos."

Two of the stars Buchler is referring to are R Scuti and AC Hercules. In 1994 Buchler and several colleagues applied their program to these stars using 15 years of observations in the case of R Scuti and 12 in the case of AC Hercules. R Scuti was shown to be described by chaotic dynamics of dimension four. The dimension of AC Hercules was less certain; it appeared to be three or four.

Long-period large red variables have recently been looked at by V. Icke, A. Frank, and A. Heske of the Netherlands. They attempted to account for irregularity (possibly chaos) of these stars by looking into how their outer layers react to pulsations that arise in a zone of instability deep inside the star. Plotting the data in phase space and examining the Poincaré section they showed that chaotic motions occur over a wide range of parameters. Their results, they claimed, compared favorably with observation.

CHAOS IN ACCRETING STARS

Pulsating stars are a natural candidate for chaos, but there is another type of star in which chaos also occurs, namely x-ray stars of irregular variability. Several objects of this type were discovered in the early 1970s when the x-ray satellite UHURU was launched. One of them was Her X-1, now known to be a binary system consisting of a neutron star surrounded by an accretion disk, and a companion star.

The signal from Her X-1 consisted of x-ray flashes of period 1.24 sec. But there were other periods associated with the object superimposed on this signal, in particular a 1.7 day period indicating that the source is moving toward us and away from us periodically. This is assumed to be caused by motion around another body—a star.

There is also evidence for an eclipse: Every 1.7 days the x rays cease for about five hours when the star apparently moves in front of the source cutting it off. Furthermore, the x rays also stop after 12 days for about 23 days.

Soon after Her X-1 was discovered astronomers began looking for its companion. About 30 years earlier a variable star called HZ Hercules had been discovered in Hercules. It was catalogued and forgotten, but in 1972, because it was so close to Her X-1, several astronomers began looking at it. Was it the companion of the x-ray source Her X-1? John and Neta Bahcall showed that its period was 1.7 days, the same as one of the periods of Her X-1. Further study showed that it is also faintest when the x rays disappeared. The evidence was overwhelming and astronomers were soon convinced that HZ Her was, indeed, the companion of Her X-1.

Because of the complexity of the variability of Her X-1 a number of people became interested in determining whether it was chaotic. A team at the Max Planck Institute for Physics and Astronomy in Germany was the first to apply the techniques of chaos to see if an attractor was present. In 1987 they reported that they had found one with a dimension between two and three; it was a fractal dimension and the attractor was therefore strange. With such a low dimension there appeared to be a good chance of modeling the system with a relatively simple model.

Their work came to the attention of Jay Norris of the Naval Research Laboratory and Terry Matilsky of Rutgers University who were also working in chaos. "We were using the same techniques as the German team when we stumbled on their paper," said Matilsky. "After reading it I realized it was something we could do, so it actually took very little time on our part to check it out. We didn't try to confirm or deny [their result] but we had all the machinery at our disposal to check it. What we ended up doing was taking a pure sine wave and adding some random noise to it, so that it had little bumps and wiggles similar to their signal and to our surprise we got exactly the same results they got. In other words, we duplicated it without any data from

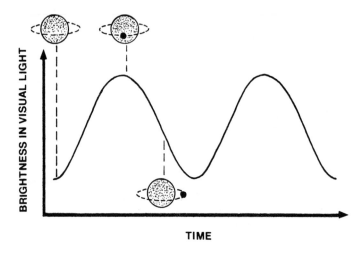

Light curve of Hercules system.

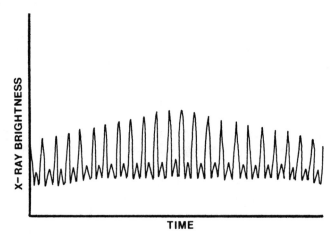

Plot of x-ray brightness versus time for Her X-1.

Terry Matilsky.

the x-ray source. So what we showed was that you have to be really careful because other types of signals mimic an attractor." In short, they showed that the German team had not found a strange attractor.

Noise is a problem in most signals. To get an idea what noise is, set your radio to a frequency that does not carry a station and turn up the volume. The hiss you hear is noise. It is produced by fluctuations in current through the electronic components of your radio. It's easy to see that noise is random; it fluctuates erratically with no pattern.

Matilsky's interest in chaos came from talks with his cousin, Mitchell Feigenbaum. "We kept in touch and he always told me what he was working on," said Matilsky. His interest in astronomy goes back to an early interest in photography. One day he saw some astronomical photographs in a magazine and decided to build a telescope to see if he could duplicate them, and indeed he did. He did his graduate work at Princeton University, sending up rockets and looking at the UV spectra of hot stars. After graduation he worked for Riccardo Giacconi of American Science and Engineering on the first x-ray satellite, UHURU. He is now at Rutgers University.

"Showing that there is not an attractor in Her X-1 was not a very romantic result," said Matilsky. "It's not like discovering there is one. Still, Her X-1 is a very interesting source and I think there are sources out there that will turn out to be strange attractors." And, of course, the presence of a strange attractor means the source is chaotic. He said that he thought the quasiperiodic oscillators (sources that are only approximately periodic) were the best candidates.

Matilsky feels that the problem with noise has to be overcome before real progress can be made in finding attractors. "I'm trying to understand the effects of noise on a signal. It's a serious problem. The theoretical models . . . Feigenbaum stuff, period doubling and so on are on a much more secure footing. They're numerical results. But when you go out and look at a signal coming from an object in the real world you have to worry about what the level of the signal is, what the noise level is and so on."

Her X-1 is not the only object Matilsky and Norris have looked at. They have also applied their program to the sources Circ X-1 and Sco X-1. In each case noise was a serious problem. But computer limitations have also been a problem. "Chaos representations take up a lot of computer time. A tremendous amount of time is required to get, say, 1000 data points, and that isn't very many," said Matilsky. To get around this, Matilsky has started to use the same approach Sussman and Wisdom used earlier; in other words, he has been designing and building his own computer. "Using computer chips called field programmable gate arrays you can do repetitive calculations very rapidly. You can get the computer power of a Cray with a chip that only costs a couple of thousand dollars," he said.

I asked him about the future role of chaos in astronomy. Will it become increasingly important? He hesitated, then replied, "It's really hard to know. There seems to be more and more things happening. People are working on a lot of problems but the big stumbling block is finding good examples of the stuff. You can predict chaos in theoretical models; they work wonderfully, but when you look for it in observational data you start to run into trouble."

"What people did in non-linear dynamics early on was approximate the complicated behavior. That was a big problem . . . with the approximations, not only was the accuracy, but the entire nature of the behavior was lost." He hesitated, then continued. "What we really need now are some new techniques. But maybe they won't come. It may turn out to be like the theory of convection; it's been bumping along for 100 years with problems that everyone knows about, but can't do anything about. Anyway, we do our best."

Another source that has attracted considerable attention in relation to chaos is the x-ray source Cyg X-1. Like Her X-1, Cyg X-1 was discovered by the x-ray satellite UHURU. Rapid flickering of its x rays, down to about 1/1000th second, indicated that it was small. Soon after its discovery astronomers began a search for a visible star in the neighborhood of the x-ray source. They quickly narrowed in on the blue giant HDE 226868; it was approximately 8000 light years from Earth and about 23 times as massive as our sun. A model was developed in which gas from the giant star was being pulled into an accretion ring around a black hole. The x rays were generated as matter from the inner edge of this ring fell into the black hole. This model, with few variations, is accepted today, and Cyg X-1 is still considered to be one of our best black hole candidates.

Once the model was established a number of people began to wonder if the dynamics in the Cyg X-1 accretion disk could be described by an attractor. One of the groups that began looking into this possibility consisted of James Lochner, Jean Swank, and A. E. Szynkewaik of Goddard Space Flight Center. Lochner worked on the project as part of his Ph.D. thesis. Using the one-dimensional x-ray light curve from Cyg X-1, they constructed the phase space and used the standard technique for searching for an attractor and determining its dimension. The dimension of an attractor is important because it indicates the minimum number of parameters necessary to describe it. It can also tell if the attractor is strange.

"What was challenging about the work was that the technique we used demanded data that were fairly clean, with low noise. But we had x-ray data from an x-ray satellite and its noise level was high. So we spent a lot of time trying to figure out how to properly account for the noise, and how it effected the result. We looked at what sort of signal was left after the noise was subtracted out and so on," said Lochner. They applied their technique to two different data sets; one from the HEAO satellite and one from the EXOSAT satellite.

Born and raised in Rochester, New York, Lochner's interest in astronomy came from reading articles and books on astronomy. "It's a bit ironic," he said, "but in my 9th grade science class I did a project on Cyg X-1. I was required to do a research project and present it to the class, and after seeing an article on black holes in *Scientific American* I decided to do it on Cyg X-1, never dreaming that one day I would be doing my Ph.D. thesis on the same object." Lockner finished his Ph.D. in 1989, went to Los Alamos for a post-doctoral, then returned to Goddard, where he still works.

Lochner, Swank, and Szymkowiak found no evidence for a low-dimensional attractor in the data, but there seemed to be some evidence of a high-dimensional attractor. I asked Lochner if there was any chance it was fractal (i.e., strange). "We couldn't tell because of the noise. We did a lot of work on the effect of the number of data points . . . you want to populate the phase space as much as possible so if there is an attractor you can see it. We had tens of thousands of data points and weren't sure at first if that was enough." They eventually decided that it was.

About the time the Goddard group published, some newer and better techniques were developed, but they have not yet been applied to Cyg X-1. Lochner also talked about a new x-ray satellite, called XTE (X-ray Timing Explorer), launched in 1995; he mentioned that it has a large detector and low noise. He expects some important results to come from it.

"What was exciting about the project, and about what the field of chaos has to offer is that you can take these seemingly complicated signals and discover things about the physical system that gives rise to the signal. That was the real attraction for me. When you start thinking and learning about accretions disks, you find they're very complicated things, and when it really gets down to it nobody really understands them. The sort of work we were doing was an attempt to decide: Is it as random as we have been assuming or is there something simpler driving the system? That's the question that is still hanging out there. I saw that the technique that nonlinear dynamics and chaos had to offer was a very promising avenue to try, an avenue that might give some answers to the problems."

Jean Swank was Lochner's thesis advisor on the project. She got her Ph.D. from Caltech in particle physics and gradually drifted into astrophysical applications of particle physics—x-ray astrophysics, in particular. She is still working in x-ray astrophysics at Goddard.

I asked her how she got interested in applying chaos to Cyg X-1. "I've always been interested in different ways of modeling the aperiodic signal from x-ray sources," she said. "I was aware of the work on chaos and nonlinear dynamics and asked myself: Is this a tool one can use to find evidence for dynamics of systems that look random?" She decided it was.

Searches for attractors in several other objects have also been made. In the late 1980s John Cannizzo and D. A. Goodings of McMaster University in Canada searched for evidence of an attractor in the light curve of the dwarf nova SS Cygni. Using data that covered a period of 21 years obtained from the American Association of Variable Star Observers, they constructed the phase space and searched for an attractor, but found no evidence for one with a low dimension.

M.J. Goupil, M. Auvergne, and A. Baglin of the Nice Observatory in France searched for evidence of an attractor in the light curve of the pulsating white dwarf PG 1351+489. They found evidence for period doubling bifurcation and some indication that the dwarf was chaotic.

Another area where chaos will likely become important in the next few years is in relation to supernovae. Work is already underway to apply chaos to the clouds of gas that come out in the explosion. Many different processes are thought to be responsible for the shaping of these clouds: turbulence, self gravity, and magnetic fields. Application of the techniques of chaos are likely to give a better understanding of these clouds.

CHAOS IN GALAXIES

So far in this chapter we have been talking about dissipative systems—systems that lose energy. They are the only systems in which attractors appear. But as we saw earlier, chaos is also important in non-dissipative systems. Hénon looked for chaotic orbits in globular clusters, and it has been known for years that galaxies—at least the part of them we see—are unstable. They are stabilized by large surrounding clouds of dark matter. Just as planets and asteroids in orbit around the sun can be chaotic, so can stars in orbit around their galaxy. Chaotic orbits have been found in several types of galaxies so it's best to begin with a brief review of the classification and structure of galaxies. A quick look at a few photographs of galaxies shows that they have many shapes; some have long dangling spiral arms, others are more tightly wound. Some are elliptical and a few appear

A galaxy. (Hale Observatories)

to be completely irregular. Edwin Hubble of Mt. Wilson Observatory set up a classification scheme in the early 1930s. He classified galaxies as spiral, barred spiral (similar to spiral except they have a bar-like structure through their center), elliptical, and irregular. He subclassified the spirals and barred spirals according to how tightly they were wound, and the ellipticals according to their shape in the sky (it varies from round to very elliptical). He didn't classify irregulars; they were put in a class by themselves.

For many years galaxies were thought to be the largest structures in the universe, but eventually astronomers began discovering that, like stars, galaxies tend to cluster. Our galaxy, the Milky Way, was shown to be part of a cluster of about 25 galaxies, which we now refer to as the Local Cluster. Then, in the 1950s, Gerard de Vaucouleurs discovered that even clusters cluster, so we have larger units that we now refer to as superclusters. Our cluster is in a supercluster known as the Virgo Supercluster, named for the huge Virgo cluster (which contains about 2500 members) that is at its center.

Besides differing in shape and structure, galaxies differ in other ways. Some are known to be strong radio sources; they are very energetic, releasing tremendous amounts of radiation and matter. Many of these galaxies appear to have explosions going on in their core; we refer to them as active or radio galaxies.

K.A. Innanen has been interested in chaos in galaxies for many years. Born in Kirkland Lake in Ontario, Canada, Innanen got his undergraduate and most of his graduate training at the University of Toronto. "I've been interested in the dynamics of the Milky Way galaxy ever since my Ph.D.," said Innanen. He became interested in the possibility of chaotic stars in our galaxy in the mid 1980s. "The bulk of stars in our galaxy follow regular orbits," he said. "If you follow the motions of stars the majority exhibit no chaotic behavior. But there is the possibility of a class of chaotic orbits which arises from low angular momentum stars, that is, stars that have a tendency to fall toward the center of the galaxy. And if there is a mass concentration at the center, and most astronomers believe there is, it's possible that when these stars approach the center their orbits will become chaotic."

What happens when a star passes close to the nucleus, according to Innanen, is that the gravitational field of the nucleus knocks it out into the outer region of the galaxy—into the galaxy's halo. He worked on the project with R. G. Carlberg who is now at the University of Toronto and N. D. Caranicolas of Greece. Their work was theoretical; they did look unsuccessfully for observational verification of their result, however. They were particularly interested in stars in the region of the sun that might become chaotic.

Hashima Hasan of the Space Telescope Institute and Colin Norman of Johns Hopkins University have also studied chaotic orbits in galaxies. Their interest was barred galaxies (galaxies with a bar-like structure through their center); they wanted to find out what effect chaos had on the bar. In their model, chaos was assumed to be the result of a huge black hole, or high mass concentration at the center of the galaxy. They examined the effect of the black hole on the orbits of stars that passed near it as the mass of the black hole was varied, and as the dimensions of the bar were varied. Examining the orbits in phase space and looking at their Poincaré section they found that the black hole (or central mass concentration) did indeed cause chaos, and that the resulting chaotic orbit would eventually cause the bar to disappear.

The transition to chaos in galaxies through period doubling bifurcation was studied by G. Contopoulos of the University of Athens. Studying two and three-dimensional models he found that as he varied various parameters the region of chaos increased. He discovered a universal constant similar to the one Feigenbaum found; it was related to the intervals between bifurcation but, strangely, it was different. He also found that stars that became chaotic in the outer regions of the galaxy could escape from it, but those that became chaotic near the core could not. Furthermore, chaos appeared to limit the elongation of the barred galaxies.

BINARY AND COLLIDING GALAXIES

In the late 1940s a radio galaxy was identified in the constellation Cygnus. Through an optical telescope it looked
strange, almost as if it were two galaxies in collision. Astronomers were excited; if true, it was our first view of colliding galaxies. In time, however, Cygnus A, as the object was later
named, was shown to be a galaxy with an exploding core, not
two galaxies in collision. Colliding galaxies were, however, soon
found. In 1966 Halton Arp of Mt. Wilson and Palomar Observatories published his *Atlas of Peculiar Galaxies,* which included
many striking photographs of colliding galaxies. Astronomers
soon became interested in simulating collisions on computers.
The first were done by Alar and Juri Toomre who were then at
New York University. They showed that filaments and bridges
could be pulled out of the galaxies as they passed in space, and
the results they got from their computer models looked amazingly like some of the colliding galaxies that were observed.
Over the years as computers have become faster and more powerful, simulations have become better.

With the disruption that occurs when galaxies collide it
seemed likely that many of the stars would end up with chaotic
orbits. Colliding galaxies therefore seemed to be a fertile ground
for searching for chaos. P. Stewart of Manchester University in
England became interested in the possibility of chaos in galactic
collisions in 1994. He set up a computer model that allowed him
to look at orbits in both binary galaxies and galaxies in collision.
In his model he released massless particles from rest in binary
systems where the two galaxies were of equal mass. He then
varied parameters and determined what conditions were needed
for chaos. As in other studies of this type he set up the phase
space, examined Poincaré sections, and determined Lyapunov
times. Considerable chaos was found in his models.

As we have seen, chaos occurs in both stars and galaxies.
Variable stars have been found that pulse chaotically, and if we
are to understand them thoroughly we must understand how

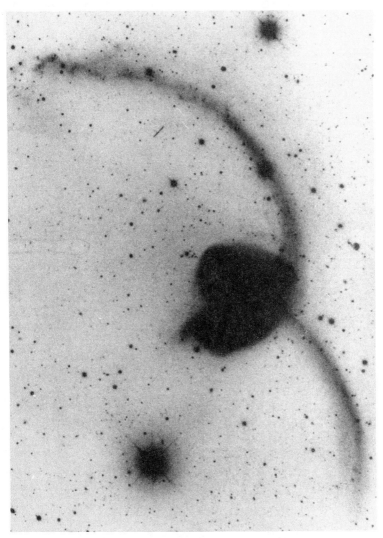

Two galaxies in collision. (F. Schweizer)

and why they become chaotic. Much has been accomplished, but it is still a fertile field. Furthermore, a number of x-ray sources are known to fluctuate erratically, and they have also been examined for chaos. Finally, there is evidence for chaotic motions in galaxies, with barred galaxies and colliding galaxies looking the most promising.

13

Chaos in General Relativity, Black Holes, and Cosmology

*E*arlier *we saw that chaos arises in phenomena described by non-linear equations. It occurs even in very simple equations such as* the one that describes the pendulum. General relativity, one of the most famous theories, is formulated in terms of a nonlinear equation. This makes us wonder if some of the phenomena described by general relativity, namely black holes, objects orbiting black holes, and even the universe itself, can become chaotic under certain circumstances. This is not an easy question to answer. The problem is the equation itself, namely the equation of general relativity; it is so complex that the most general solution has never been obtained. It has, of course, been solved for many simple systems; if the system has considerable symmetry (e.g., it is spherical) the equation reduces to a number of ordinary equations that can be solved, but chaos does not occur in these cases. In more realistic cases—situations that actually occur in nature—chaos may occur, but the equations are either unsolvable or very difficult to solve. This presents a dilemma. If we try to model the system using many simplifications it won't exhibit chaos, but if we model it realistically we can't solve it.

GENERAL RELATIVITY

Einstein introduced his special theory of relativity in 1905. It was a new and different theory, a theory that attacked the traditional view of space and time. Since Newton's time, both space and time had been accepted as absolute. In other words, they were assumed to be the same for all observers throughout the universe. Einstein's theory challenged and broke down these "pillars" of early science, and in so doing introduced a number of seemingly strange ideas. They were so foreign to the scientists of the time that, despite the evidence that they were correct, it took years for them to be accepted. According to Einstein's theory, the rate at which the clock of one observer ran relative to that of a second observer depended on how fast they were moving relative to one another, in other words, their relative velocities. Scientists found this hard to accept; it didn't seem possible. Objects also contracted in the direction of their motion according to Einstein's theory. According to his measurement, a yardstick traveling at close to the speed of light relative to an observer would only be a few inches long.

Special relativity was a hard pill to swallow for many. It was different, and to many, bizarre. But within a few years it was acknowledged as the correct view of nature—a milestone of science, one of the greatest inventions of the human mind. Einstein was indifferent to the world-wide fame that came to him when the theory was accepted. His mind was on other things. He realized that the theory was incomplete: It applied only to straight-line uniform motion and said nothing about accelerated motion, which in the world of physics was equally as important as uniform motion. Therefore, Einstein set out to generalize the theory.

Out of Einstein's investigation came what is called the principle of equivalence, a postulate that assumes an equivalence between acceleration and gravity. If you were in an elevator that was accelerated upward at, say, 32 ft/sec^2 (the acceleration of gravity here on Earth), you would feel as if you were standing on the surface of the Earth. In fact, there was no way you could prove you weren't.

Einstein struggled with his ideas for about ten years. Finally, in 1915, everything came together and he arrived at a set of equations that we now refer to as the field equations of general relativity. Interestingly, the theory turned out to be more than a simple generalization of special relativity; because of the equivalence of gravity and acceleration, it was a theory of gravity. But Newton had formulated a theory of gravitation 300 years earlier, and it was acknowledged to be an excellent theory. In fact, no one had ever found any flaws in it. What good would a theory be that gave exactly the same results? This, it turned out, wasn't a problem. Einstein's theory not only predicted everything Newton's theory did, it predicted more. Furthermore it was based on completely different ideas and concepts. Newton had thought of gravity as an action-at-a-distance force across space—a mysterious force that was inexplicable. Einstein interpreted gravity as a curvature of space, a curvature we could not see but could describe by equations. The planets moved through this curved space along the shortest possible trajectories, what we refer to mathematically as geodesics. You might think a geodesic can only be a straight line, but this isn't always the case, even in the limited world of our experience. A geodesic on the surface of the Earth, for example, is the arc of a circle. In three-dimensional space, or more exactly, in four-dimensional space-time, a geodesic can be a curved path, and indeed for the planets in orbit around the sun, it is.

As a first condition, Einstein's theory had to give Newton's equation, and indeed as Einstein showed, it did. Einstein not only got Newton's equation, but he got it with an extra term. What was the significance of this term? Einstein showed that it predicted a slow movement of the major axis of the elliptical orbit, what we call precession. Astronomers had noticed a deviation from the predicted orbit of Mercury, but for years had assumed that it was caused by a planet inside Mercury's orbit.

When Einstein's equations were used to calculate Mercury's orbit, the deviation was completely accounted for.

The theory also predicted that a light beam passing through a gravitational field would be deflected; the image of a star passing near the sun would therefore appear closer to the sun than it really is. This was verified in a 1921 eclipse. Finally, it predicted that gravitational fields affect the rate at which clocks run. A clock in a strong gravitational field would run slower than one in a weak field; the stronger the field, the bigger the difference. And this was also verified.

Our interest, however, is in the nonlinearity of the field equations and the chaos that can result because of this nonlinearity.

THE NONLINEARITY OF EINSTEIN'S EQUATIONS

We talked briefly about nonlinearity in an earlier chapter but it is useful to take another look at it in relation to Einstein's equations. Nonlinear equations have many properties and difficulties that linear equations do not have. If a nonlinear equation, for example, describes a collection of objects and we want to find the collective effect of these objects, we cannot merely add their individual effects. Because source and effect are independent of one another, their sum does not give the overall effect. With nonlinear phenomena there are strong interactions between the bodies and the contribution from each is modified by the others. Mathematically this means if we change a variable on one side of the equation it doesn't cause a proportional change in the variable on the other side.

Interestingly, for most observable phenomena involving the gravitational field, the nonlinearity of Einstein's equations is not important. It is possible to approximate the equations by a set of linear ones and in most cases they are sufficient. Among the three tests mentioned earlier, for example, only the precession of Mercury's orbit is strongly nonlinear.

Still, nonlinearity may play an important role in the universe, so it is critical that we examine it. To see the full effect of the nonlinearity we have to go to places where the gravitational field is extremely strong. When a star collapses, for example, its gravitational field increases, so the end state—objects such as black holes and wormholes in space—is strongly nonlinear. Nonlinearity is also important in the early universe. According to the big bang theory, our universe was created from a singularity—an infinitely dense point—approximately 15 billion years ago, and it is possible that it will eventually collapse back on itself in what is called the "big crunch."

Nonlinearity is important because it can lead to chaos. That's not to say that we get chaos all the time with nonlinear equations; in practice it only occurs under certain conditions. Einstein's equations, for example, do not exhibit chaos when there is a lot of symmetry in the system.

The nonlinearity of Einstein's equations is also important in another problem. One of humanity's aims has been to find a theory that explains everything—a unified theory of nature, or theory of everything. Einstein spent the last 30 years of his life searching for a limited version of such a theory. He wanted to unify the gravitational and electromagnetic fields into a single theory—a generalization of his general theory of relativity—but he didn't succeed, and no one has succeeded since. In fact the problem has become more complicated now as two other fields of nature are now known and they would have to be included for a complete theory.

The reason general relativity cannot be unified with electromagnetic theory seems to be related to its nonlinearity. To unify the two fields properly we have to construct a "quantized" version of relativity; in other words, we have to quantize it, and so far no one has. It is a problem that has frustrated generations of physicists. We have been able to quantize the theory of the electromagnetic field, but the equations in this case are linear. Scientists have generally given up trying to quantize gravity, and are now trying a different approach to the problem.

One of the major problems with general relativity is that it is not a theory in the usual sense. In the case of most theories you have a stable background, or frame of reference, and you look for solutions within it. In general relativity the solution is the background—the space-time—and it is not necessarily stable. As we saw earlier neither space not time are absolute as they are in Newton's theory. Space can become warped and time depends on the warpage. "Einstein's theory . . . does not provide us with a set of parameters evolving in time in the way classical theory does," writes Svend Rugh of the Niels Bohr Institute of Copenhagen, Denmark. Rugh has been working on the problem for several years. He would like to develop a better definition of chaos in general relativity, one that gets around the above problems.

He is also concerned with the convergence of trajectories as a measure of chaos. Einstein's equations are like any other equations in that if the initial conditions are specified they determine the evolution of the system. In other words, they tell us what happens to trajectories—whether they stay together or diverge, and if they diverge, whether they diverge fast enough to be chaotic. But there is some ambiguity according to Rugh. "What do we really mean by 'nearby' trajectories in general relativity?" he asked. "It is not clearly defined."

Let's take a closer look at the solution to Einstein's equations. Einstein, interestingly, was not the first to obtain a solution. It came from Karl Schwarzschild, a soldier in the German army who was stationed on the Russian front in World War I. Schwarzschild had come down with a rare disease and was bedridden when he saw Einstein's paper, but this didn't stop him. Within a few days he had obtained a solution which he sent to Einstein.

Einstein was surprised and delighted that a solution had been found so quickly. He wrote back to Schwarzschild telling him he would present it at the next meeting of the Prussian Academy. Schwarzschild, unfortunately, never lived long enough to see the fruits of his effort; he died a few months later.

The solution obtained by Schwarzschild was one of the simplest. It was for a sphere, but because most objects in the universe are spherical, it was a particularly important solution. We now know that unless there is an outside influence, spherical objects do not give rise to chaos, so Schwarzschild's solution by itself is not important in relation to chaos. Since Schwarzschild's time, however, many solutions for Einstein's equations have been obtained (solutions for systems that are not spherical) and chaos may be important in relation to some of them.

THE BLACK HOLE

One of the most important predictions to come out of the equations of general relativity is that of a black hole. To Einstein and others who saw the seeds of a black hole in the equations, the prediction was unsettling. To them it seemed more like a deficiency of the equations, rather than something new and exciting.

The black hole concept, strangely enough, had been around for over a hundred years before it was predicted from relativity. In 1874 John Michell of England showed that if the mass of a "star" was sufficiently great it could trap its own radiation and would become invisible. Laplace, the great French mathematician, arrived at the same conclusion a few years later; he discussed the possibility in one of his books, but became embarrassed by the idea and omitted it in the second edition of his book.

To understand black holes, how they form, and why they are exotic, we have to start with what is called escape velocity— the velocity needed to completely escape a gravitational field. The escape velocity of the Earth is about 25,000 miles per hour, which means that if a rocket blasts off from the surface of Earth with this velocity it will not go into orbit, but will escape to space. Escape velocity depends on mass, so planets with greater masses will have higher escape velocities, and stars and other objects which are even more massive will have exceedingly high escape velocities. But according to relativity there is a limiting velocity in the universe—the velocity of light. What happens when the escape velocity is greater than this? Is this possible? We can, indeed, visualize, as Michell and Laplace did, the Earth expanding at constant density until it has an escape velocity equal to the velocity of light; both men showed that this would happen when it was as big as the orbit of Mars. Just beyond this it would not be visible because photons (particles of light), which travel only at the speed of light, could not leave it.

Do such objects exist? We don't know of any planets such as this, but the same phenomenon occurs when a star begins to run out of fuel and collapses in on itself. Throughout most of its life a star is balanced between two forces—an inward gravitational force and an outward force due to the gas pressure—but as the star ages the nuclear furnace at its center finally flickers and dies, and the star collapses. Robert Oppenheimer and his students showed that the collapse can be catastrophic; all of the matter of the star can collapse to a point called a singularity. Strangely, though, this singularity is surrounded by a black spherical surface, a surface we now call the event horizon. It is called an event horizon because it represents the end of events associated with the universe; once you cross it you can never get back through it to the outside world.

Black holes, as these objects are called, have many strange and bizarre properties. If you approached one, tidal forces would pull you apart, and if you compared the time on your watch to that of a distant observer, you would notice that it was different. You would see the distant observer's watch running fast; he would see yours running slow. Yet, to you, time would appear to pass normally.

Interestingly, black holes have only three possible properties: mass, charge, and spin. This allows only four types of black holes: the Schwarzschild black hole that has only mass, the Reissner–Nordström black hole that has charge, the Kerr black hole that has only spin, and the most complex, the Kerr–Newman black hole that has mass, spin, and charge.

What a black hole might look like in space.

So far we've talked only about what happens according to Einstein's theory, but what about observation? Have we identified any good black hole candidates? Indeed we have, but the number is much smaller than most astronomers would like. No one is sure why, but it takes a particularly massive star—about 20 solar masses—to produce a black hole (the final mass after it is formed, however, need only be greater than 3.2 solar masses), and this may be the main problem. One of the oldest and best known candidates is Cyg X-1, in the constellation Cygnus. It is an x-ray system that pulses rapidly; the shortest pulses indicate the source of the x rays is exceedingly small, not over a few miles across, and therefore too small to be seen with a telescope. The x rays are caused when matter from a nearby star (the primary) is pulled into a small collapsed object (the secondary) which may be a black hole. When the primary was identified through a telescope, it was found to be a large blue star. With the mass of the primary known (and making various assumptions) the secondary was shown to have a mass of about eight suns, easily massive enough to be a black hole.

BLACK HOLES AND CHAOS

Black holes are exotic, but of more importance to us, they are described by the nonlinear equations of general relativity and are therefore good candidates for chaos. Chaos can occur in black hole systems in two ways. The first, which is generally referred to as geodesic chaos, is the chaos we saw earlier in connection with the planets and asteroids. If a small object, a planet for example, were orbiting the black hole, its orbit could become chaotic if appropriate resonances were present. In general, however, we do not get chaotic orbits in this case. Even if the black hole is complex, a Kerr–Newman black hole for example, chaos is unlikely. An external perturbation of some type is needed; it can be supplied by another object, or by immersing the black hole in a magnetic or electromagnetic field.

The chaos described above occurs only in nondissipative systems, but dissipative systems involving black holes are also possible, and they can also give rise to chaos. This is the second type referred to above. Consider two black holes, or a black hole and a neutron star, orbiting one another. Considerable radiation, both electromagnetic and gravitational, would be given off in such a system, and it would therefore be dissipative. The presence of a strange attractor in such a system would signify chaos.

Luca Bombelli, who is now at Mercyhurst College in Erie, Pennsylvania, and Esteban Calzetta of the Institute for Astronomy in Buenos Aires, Argentina, have been studying a system consisting of a particle (representing, perhaps, a planet) orbiting an ordinary spherical, or Schwarzschild, black hole that is perturbed in some way. The perturbation could be caused by a small object, for example.

Born in Switzerland to an Italian father and a Swedish mother, Bombelli grew up in Spain and went to college in Italy. His interest in astronomy and mathematics began early. "I always liked mathematics, even in the lower grades," he said.

He came to Syracuse University in the United States to do graduate work in experimental physics, but began attending some of the talks given by the relativity group. At these meetings he met Peter Bergmann, a former collaborator of Einstein's. "The relativity group was so good I switched to relativity," said Bombelli. "Working with that group was a real inspiration."

After his Ph.D. he went on several post-doctorals, one in Belgium at the Free French University of Brussels. Several groups were studying chaos, and although he was in the relativity group, he began attending some of their talks. It was during this time that Esteban Calzetta came from Argentina as a visiting professor; Bombelli met him, and soon the two men were working together.

"A lot of things were known about chaos," said Bombelli, "but I wasn't sure what they really meant physically. We started looking for a system that was relevant to astrophysics and decided to look at black holes." They set up a computer program for a Schwarzschild black hole which was orbited by a test particle.

One of the difficulties in any system of this type is identifying chaos when it occurs. As we saw earlier, chaos is associated with exponential divergence of trajectories in phase space, so it was natural to examine this divergence as the system evolves. Is the divergence rapid enough? This is usually determined by calculating the Lyapunov exponent we talked about earlier. But there are serious difficulties with this method and Bombelli and Calzetta decided to use a different method. Called the Melnikov method after its discoverer, it is a search for a Smale horseshoe within the data. As we saw earlier Smale showed that chaos is associated with a stretching and bending of phase space—a process that creates a horseshoe-like figure; the Melnikov method gives a procedure for looking for the horseshoe. Using this method, Bombelli and Calzetta found evidence for chaos in the orbits.

Bombelli and Calzetta are now several thousand miles apart, but they are still working together, and they have turned to a different problem. "The system we're interested in now is not a geodesic system," said Bombelli. "We're interested in a two body problem, either two black holes revolving around one another, or a black hole and a neutron star. Such a system would give off gravitational radiation, and would be dissipative." Astronomers have found a system of this type called the binary pulsar that is believed to consist of a black hole and a neutron star. Much has been learned from it, but so far no one has looked for chaos in its motion. The system is dissipative, so Bombelli and Calzetta will be looking for evidence of a strange attractor. It is a much more complex problem than the one they worked on earlier. All of the work on the previous problem was analytical, but according to Bombelli they will be using both analytical and numerical (computer) methods on this system.

For chaos to occur in single black holes an external perturbation is needed, but when a particle orbits two black holes

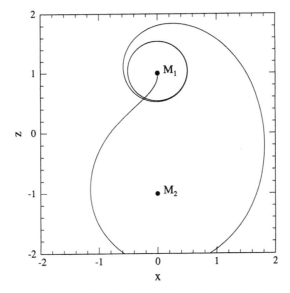

Photon orbits around two black holes M_1 and M_2. (Matthew Collier)

things are quite different. The geodesic problem for two black holes was solved by Nobel Laureate Subrahmanyan Chandrasekhar of the University of Chicago in 1989. Chandrasekhar is well known for his work on black holes. His book, *The Mathematical Theory of Black Holes*, is the bible of the industry.

Chandrasekhar found that some of the orbits in the double black hole problem were puzzling, so he got in touch with George Contoupolos of the University of Athens, an acknowledged expert on chaos who had published several papers in the area, and asked him to check them out. Although centered at the University of Athens, Contoupolos has, in recent years, spent part of his year at the University of Florida. Setting up the problem he found that several types of orbits were possible around the black holes. He considered the orbits of both light particles—photons—and matter particles; some of them were

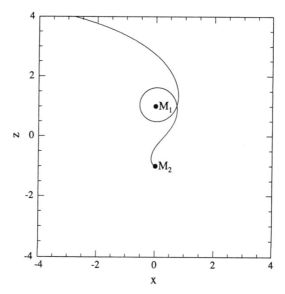

Photon trajectories near two black holes. Photon disappears into black hole M_2. (Matthew Collier)

bound (didn't fly off to space), others were not. The bound ones were, of course, the most interesting. Among them he found several types of orbits; particles could orbit both black holes or make a figure eight around them. In some cases the particle eventually fell into one of the black holes.

Contoupolos showed that the orbits of the photons were completely chaotic, and most of the orbits of the matter particles close to the black holes were chaotic, but some were not. In particular, he showed that between any two orbits of different kinds there was an orbit of a third kind. This is a Cantor set, and tells us that chaos is present.

The motion of particles around black holes immersed in a magnetic field has been investigated by V. Karas and D. Vokrouhlicky of Czechoslovakia. Plotting the orbits in phase space, they looked at Poincaré sections and calculated Lyapunov expo-

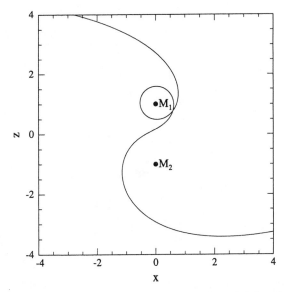

Photon trajectory near two black holes. Photon circles first black hole and escapes to space. (Matthew Collier)

nents and reported that they detected chaotic motions. A similar calculation was made by Y. Nakamura and T. Ishhizuka of Japan, who also reported chaos for this case.

The evidence for chaos in systems involving black holes is now overwhelming, and many important discoveries will no doubt be made in the area in the next few years.

COSMOLOGY AND CHAOS

One of the most important applications of general relativity is to the entire universe, and Einstein began working on this problem shortly after he finished his theory. He soon discovered, however, that the universe described by his equations was unstable; it tended to expand or contract, and Einstein, who had been assured by astronomers that the universe was stable, wanted it to be held fixed. To stabilize it he introduced a term called the "cosmological constant." It was important on a very large scale—on the scale of the universe—but didn't effect things on a small scale, and it kept his universe from expanding.

In the same year that Einstein published his cosmology, a Dutch astronomer, Willem de Sitter, also published one. In looking at the equations of Einstein's model, de Sitter noticed a solution that Einstein had missed. De Sitter's model soon became a serious rival to Einstein's model despite the fact it appeared to have a serious flaw: it was empty. De Sitter apparently didn't worry about this. "After all," he said, "our universe is nearly empty." Later it was discovered that if you put two bodies in his universe, they separated; in effect, the universe described by his equations expanded.

For years there was considerable controversy over which model was correct. Then in the late 1920s and early 1930s Edwin Hubble showed, using the giant telescopes of Mt Wilson, that all galaxies were moving away from one another; the universe was expanding. As a result both theories soon fell into disfavor.

With the discovery of the expansion of the universe another theory came to the forefront. A Russian mathematician, Aleksandr Friedmann, had looked at Einstein's equations and decided to follow up on the case Einstein had discarded—the model with no cosmological constant. Developing the resulting equations, Friedmann found that there were three possible universes within his model: a positively curved universe (like the surface of a ball) that expanded to a certain radius, then collapsed back on itself, a negatively curved one (like the surface of a saddle) that expanded forever, and a flat one that also expanded forever. Friedmann showed that there is a critical mass density. Over this density the universe is closed (collapses) and is positively curved; under it, it is open and negatively curved.

Friedmann's model soon became the accepted one. A streamlined version was published by Robertson and Walker in the United States, and it was eventually called the Friedmann–Robertson–Walker model. Abbe LeMaître of Belgium later made important contributions to it.

Other models are also possible within the Einstein framework. One that has attracted a lot of attention in relation to chaos is called the Mixmaster model. It is not a realistic model but has a number of interesting features that relate to the early universe. The model was formulated in 1969 by Charles Misner of the University of Maryland at College Park. As the name suggests, the universe it describes undergoes mixer action, like a kitchen mixer: it expands in two directions and contracts in the third. Viewed from the outside it would appear to alternately flatten into a pancake and stretch into something that looks like a cigar.

The Mixmaster model (named after one of the brands of mixers) is important for several reasons. First, it is a relatively simple model, and for the study of chaos, simple models are required. It is extremely difficult to study chaos using the full Einstein equations; they are much too complex. What is needed are special cases, and the Mixmaster model fills the bill. Furthermore, it has been thoroughly studied for several decades, and is fairly well understood. It is a homogeneous model in that it is the same everywhere, and the resulting Einstein's equations can be reduced to a set of ordinary differential equations that are relatively easy to solve.

Of more interest to astronomers, however—particularly in light of recent discoveries about the large-scale structure of the universe—are inhomogeneous cosmologies, and there are several indications that the Mixmaster model is related to them. In the early 1970s three Russians, V. A. Belinski, I. M. Khalatnikov, and E. M. Lifshitz decided to look at inhomogeneous solutions to Einstein's equations. Restricting themselves to the very early universe (more exactly, the singularity), they found indications that Mixmaster action would occur.

Andrew Zardeki of the Los Alamos National Laboratory followed up on this work in 1983 by constructing numerical solutions to the equations. His computations showed that the universe oscillated as it emerged out of a singularity in the early universe, and when he calculated the Lyapunov exponent he found it to be positive, indicating the oscillations were chaotic (this also applies to the last stage—the singularity of the "big crunch" in the collapsing universe).

David Hobill, who was then at the National Center for Supercomputing Application in Champaign, Illinois, read Zardeki's paper and was puzzled. Born in Massachusetts, Hobill got his undergraduate degree from Worcester Polytechnic and his Ph.D. from the University of Victoria in Canada, where he did a thesis on gravitational waves. After graduation he worked on collisions of black holes. While working on them he started reading about chaos and became interested in it.

"When I looked at the graphs Zardeki had made they appeared to contradict the approximate results the Russians had obtained," said Hobill. "Zardeki said that as you approach the Big Bang or Big Crunch singularity that the results should approach those of the Mixmaster model. That means there should be two expanding directions and a collapsing direction. But his models sometimes had three expanding directions and three collapsing directions." This prompted Hobill and three colleagues to write their own computer program to try to duplicate the results, but when they ran it they found they didn't get Zardeki's results; their results, however, were consistent with the Russian's calculations. Hobill decided that Zardeki's numerical method had to be at fault, so he wrote a computer program using the same technique Zardeki had used, and lo and behold he got Zardeki's results.

Something was obviously wrong! Looking carefully through what he had done, Hobill saw that some of the constraints of the theory were not satisfied when Zardeki's method was used. "If these constraints aren't obeyed you know the full set of Einstein equations are not satisfied," said Hobill. Looking further, Hobill saw that in neglecting the constraints Zardeki was, in effect, introducing a negative energy, and it was this negative energy that was causing the chaotic oscillations. So far there is no proof that negative energies exist, but as we will see later, negative energies are associated with an inflation that may have occurred early on in the universe.

Hobill went on to examine the results further. Like Zardeki, he and his colleagues calculated the Lyapunov exponent and showed that it was zero. But strangely, he found that it wasn't zero if he didn't take the limit as time went to infinity (as is required by the definition). In other words, a "weakened" definition of the exponent wasn't zero. "My point of view is that it's definitely not chaotic from the definition of strong chaos. But when you weaken the definition it has weakened chaos. I think everyone working in the area agrees what is going on, but no one agrees on how to define what is going on."

Two other groups also noticed there were problems with Zardeki's results at about the same time Hobill and his colleagues noticed them.

Hobill came to the University of Calgary after his post-doctoral in Illinois, where he continued to follow up on the problem, but when he found that most of the Mixmaster oscillations occurred when the universe was microscopic in size—so small it couldn't be properly described by general relativity—his enthusiasm began to wane. A quantum version of general relativity was needed and it wasn't available.

"We were extending the solutions to Einstein's equation beyond where they were physically valid," said Hobill, "so we weren't really going to find out anything about the physics of the early universe."

As Hobill pointed out, however, even though the model is not officially chaotic, information (position, momentum) is still being lost. It's not being lost at a rate fast enough to qualify as chaos. Nevertheless if you start with a finite amount of information about a system, all of it will eventually be lost.

One of the major problems in a system of this type is how to qualify chaos. It can be defined in several ways, but each has its difficulties. A positive Lyapunov exponent signifies chaos, as does the presence of a strange attractor, a fractal

A plot showing chaotic oscillations in the Mixmaster universe. (David Hobill)

dimension, or the presence of a Cantor set. Hobill is presently trying to apply some of the other definitions to the Mixmaster model to see what they give. His major interest now, however, is inhomogeneous models. Most of the work that has been done so far has been on homogeneous models, but according to Hobill, inhomogeneous models are much more interesting, and likely more realistic.

Beverly Berger of Oakland University in Rochester, Minnesota, has also recently become interested in inhomogeneous models. Born in Paterson, N.J., Berger got her bachelor's degree from Rochester University, then went to the University of Maryland for her Ph.D. She learned about the Mixmaster model early on, as her thesis advisor Charles Misner was the inventor of the model.

Berger became interested in the problem when she saw the papers by Hobill and others contradicting Zardeki's result. She wondered what was going on, and decided to find out for herself.

"I wanted to understand the relationship between the numerical solutions (that Hobill and others had obtained) and the approximate analytical solutions of the Russians," said Berger. Just as Hobill had earlier, she found the Lyapunov exponent was zero, but putting in a cutoff (in other words, she didn't take the limit as time went to infinity, but allowed it to be arbitrarily large) she found it was positive. "The question then was: Is it chaotic or not? My point of view is why worry about what you call it as long as you know what is going on."

Berger is presently working with Vincent Moncrief of Yale University on inhomogeneous models. They are looking into the Russian's conjecture that inhomogeneous cosmologies show Mixmaster behavior near the singularity by solving Einstein's equations numerically using supercomputers. This might be important in relation to the formation of galaxies and the large scale structure of the universe. Astronomers are still troubled over how the large-scale inhomogeneities arose. An interesting possibility is that they are inherent in the Einstein equations.

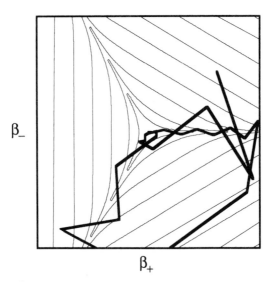

Trajectory of a Mixmaster universe in a superspace. Plotted by Beverly Berger.

"The project involves numerically studying the development of singularities in the solutions to Einstein's equations," said Moncrief. "It is not necessarily aimed at chaos, but chaos may come out of it." He paused. "I don't really expect Mixmaster behavior or chaos to occur except in special cases, but we don't have an alternative conjecture. We just expect all hell to break loose as we get close to the singularity."

Matthew Choptuik of the University of Texas at Austin is working on a similar problem. He's not interested in the collapse of the entire universe but rather a small section of it. He's trying to determine what happens when radiation collapses to form a black hole. Born in Manitoba, Canada, Choptuik obtained his undergraduate degree from Brandon University and his Ph.D. from the University of B.C.

"I wasn't really looking for chaos," he said. "I was looking at the spherically symmetric collapse of a mass where the energy is purely kinetic."

Whether a black hole forms or not depends on the intensity, but not the shape, of the initial light pulse. There is, in effect, a threshold. "The game is to look at what happens at the threshold for black hole formation," said Choptuik. "Surprisingly, we found a unique solution that has echoing properties. The system is not chaotic, but there is nonlinear behavior that has some of the features of chaos."

J. Wainwright of the University of Waterloo in Canada is also working on Mixmaster models. "My research has been to formulate the Einstein field equations as a dynamical system," said Wainwright. "The idea is to try to describe the oscillatory behavior in the past by means of an attractor. I've been able to describe what the attractor is geometrically. It's not a strange attractor; it's a normal one." He interprets the model as being weakly chaotic.

Finally, chaos has also been found in the Friedmann–Robertson–Walker cosmology when it is coupled to another field. Esteban Calzetta and Claudio El Hasi of Argentina showed that chaos can occur in such a system. They stated that this chaos sets strong limitations on our ability to predict the field at the big crunch from a given value at the big bang.

The study of chaos in general relativity and cosmology is still in its infancy, but there are indications it plays an important role, and a number of physicists are working hard to uncover its implications. Considerable progress has been made and much more is likely to come in the future.

14

Quantum Chaos and the Early Universe

*S*o far our discussion has been restricted to classical chaos—chaos *that occurs in the classical (Newtonian or relativistic) description* of physical systems. By the mid 1800s, however, scientists had begun to realize that the answers that Newtonian mechanics gave didn't always agree with observation; in some cases there were serious discrepancies. It could, for example, give the trajectory of a billiard ball bouncing around on a billiard table, but not the radiation curve of a heated object. A new approach was needed, particularly at the microscopic level, and in the 1920s it came in the form of quantum mechanics.

But if chaos occurs in the world of classical mechanics, isn't it possible that it also occurs in the world of quantum mechanics? We know that the planets and asteroids of the solar system can become chaotic, and we can easily visualize the solar system becoming microscopic in size. In fact, from a simple point of view, the atom can be thought of as a miniature solar system.

Also, we know that vibrating systems such as pendulums, vibrating springs, and rods can become chaotic, and we have oscillations at the atomic level. Can they become chaotic? Before we look into this we have to consider another problem. We know that chaos implies unpredictability, and quantum mechanics has a form of unpredictability within it already. It is contained in what is called the uncertainty principle. According to this principle, we cannot predict the orbit of an electron in an atom accurately; we can only give its probable position. Furthermore, we can't predict exactly when a radioactive nucleus will decay; we can only give the probability that it will decay within a certain time. So we have to be able to distinguish the unpredictability of chaos from that inherent in quantum mechanics.

Also, quantum theory is only important on the atomic scale. Quantum chaos, assuming it exists, would therefore seem to be of little importance in astronomy. This is not the case, however. Many of the processes studied in astronomy are described by quantum mechanics. The light that astronomers analyze, for example, depends on our knowledge of quantum mechanics. Furthermore, on the very largest scales—in the study of cosmology and the structure of the universe—quantum mechanics is important. Astronomers are still not certain what caused the large structure we see, but it can be traced back to the early universe, a universe that was microscopic in size, and because of this, quantum mechanics is crucial to our understanding of this structure.

QUANTUM THEORY

The roots of quantum theory go back to the year 1900. One of the fundamental problems in physics at the time was the emission of radiation from a heated object. A metal, when heated, gives off more radiation at certain frequencies than others. For an idealized object called a black body, a plot of radiation intensity versus frequency has a characteristic curve. At a given temperature, it rose as you went to higher frequencies, peaked, then fell off. At a different temperature it peaked at a different frequency. You can easily see this by heating an iron ball: As its temperature rises it turns red, then orange, and finally blue-white.

It was an intriguing phenomenon, but no one could explain what was going on. Physicists searched for an explanation—for

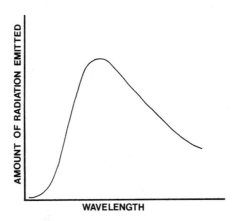

Radiation curve. A plot of the amount of radiation emitted at various wavelengths.

a mathematical formula that would fit the curve—but the best they could do was fit part of it. The problem was the turnover—the peak. None of the formulas predicted it.

Then came Max Planck, a young German professor at the University of Berlin. Realizing that traditional approaches wouldn't work, Planck tried something completely different. According to the accepted dogma, radiation was emitted continuously. Planck made the assumption that it was given off in chunks—what he referred to as quanta. To make his curve fit the experimental one, however, he had to assume the existence of a new constant that he called h, and it made him suspicious of the approach.

But it worked; Planck's formula fitted the experimental curve perfectly. Although he didn't realize it at the time, his new approach was a dramatic breakthrough in physics, a breakthrough that would lead to a revolution in science.

Scientists soon saw that Planck's constant was related to the structure of atoms. Rutherford had shown that the atom was like a miniature solar system with the sun replaced by a nucleus and the planets by electrons. The Danish physicist Niels Bohr became intrigued with the new quantum ideas and decided to apply them to Rutherford's model. To his delight he found that he could explain the spectral lines of the hydrogen atom (they can be seen by passing the light from heated hydrogen through a spectroscope) if he assumed the electrons were restricted to specific orbits specified by Planck's constant. When they jumped between orbits they emitted or absorbed photons, and it was these photons that gave rise to the spectral lines. Bohr's formula worked well with hydrogen. It predicted the position of all the lines. Strangely, when he applied it to helium it fell apart; the predicted values did not agree with the observed lines.

Bohr had obviously taken a step in the right direction, but it soon became obvious that a deeper understanding was needed to explain why his method worked. And it came within a few years from a most unlikely source, a French prince, Louis de Broglie. It was well-known at the time that light acted like a wave. It had a wave motion associated with it. De Broglie hypothesized that particles of matter also had waves associated with them. In particular, the electron in Bohr's hydrogen atom had a standing wave associated with it. This is the type of wave you get when you shake a rope so that the humps that form in it remain stationary (they don't move along the rope). According to de Broglie each orbit of the hydrogen atom had a different number of humps around it, but always an integral number.

A standing wave of the type obtained by de Broglie.

Scientists were skeptical of de Broglie's idea, but when Einstein took an interest they knew it had to be important. The following year experimental evidence for the waves was found by Clinton Davisson of Bell Laboratories in the United States.

Word spread fast, and soon scientists throughout Europe were anxious to learn about the new theory. Erwin Schrödinger, a professor at the University of Zurich, was asked by Peter Debye to give a colloquium on it. Schrödinger accepted, and at the next colloquium gave a clear account of de Broglie's work, describing his standing waves and how they accounted for Bohr's orbits. When he finished his talk Debye stood up and said, "Schrödinger . . . you are talking about waves but where is your wave equation?" The remark stuck with Schrödinger, and he began to think about it. It was true: A wave equation was needed, but how would he represent the wave? Schrödinger decided to use a "wave function," which he designated by the Greek letter psi. He derived an equation for it, an equation that is now one of the most famous equations in science.

Schrödinger did more, however, than write down the equation. He gave us the foundations of a new description of the atom and light based on waves, a theory that is sometimes referred to as wave mechanics. Furthermore, over the next few months he used it to solve some of the most difficult problems of physics.

But what was the exact nature of the wave function? No one, including Schrödinger, was sure. Schrödinger was convinced it somehow represented the electron as a "wave packet," but he discovered that this packet wouldn't remain confined; it spread rapidly over space. In 1926 Max Born of Göttingen proposed that the wave function didn't represent the electron itself, but was a "probability" wave. It gave the probability of finding the electron at a certain position. This is the interpretation we now accept.

About the same time Schrödinger published his theory, a theory based on completely different concepts (it used arrays of numbers called matrices) was put forward by Werner Heisenberg of Germany, and Heisenberg's theory gave the same results as Schrödinger's. It seemed strange that two apparently different theories gave the same results. What did it mean? At first no one was sure, then Schrödinger showed that the two theories were equivalent—just two different versions of the same theory.

By the late 1920s quantum mechanics was well-established and had been applied to many different problems. It was an excellent theory and worked well in the realm of atoms, yet it was radically different from classical mechanics. Unlike classical mechanics it restricted particles to a spectrum of energy levels. Most of the time the particles were in their lowest, or ground, state but if you shone a light on them, they would jump to higher energy levels, or excited states.

QUANTUM CHAOS

We cannot see the world of quantum physics, but through experiment we know the results that come from applying the equations of quantum theory are excellent. But the results that come from classical mechanics in the macroscopic world are also excellent. This means that somewhere the two theories have to come together; in other words, over a small range they have to give the same answer. Niels Bohr realized this and formulated it as a principle—what is now referred to as the correspondence principle. It states that in the limit, when wave effects become negligible, the two theories have to give the same result.

One of the best places to see the transition is in the hydrogen atom. When the energy of the electron is low the energy levels of the hydrogen atom are widely separated. But if you shine light of the appropriate frequency on the electron it will be kicked up to a higher energy level, and the levels here are closer together. In fact as you go to higher and higher levels they continue to get closer together until they finally merge into a continuum. This is where the transition occurs. Classical mechanics says the electron can have a continuum of energies, and indeed the highest levels are a continuum.

The question that now comes to mind is: what if the classical system is chaotic? Will it remain chaotic when it passes through the transition to the quantum realm? And if not, what happens? We see immediately that there is a problem. Classically, a system can be chaotic only if it is described by a non-linear equation, but once it has made the transition to the quantum realm it is described by Schrödinger's equation which is linear. Chaos would therefore appear to be impossible in the quantum realm. Does the system shed its chaos as it passes through the transition?

Around 1980 a number of investigators began to look into this problem. The investigations were theoretical, and only approximate models were used. Nevertheless, they gave considerable insight into the problem. Involved in the study was Joseph Ford of Atlanta, Georgia, Boris Chirikov and Felix Izraelev of the USSR, and Giulio Casati of Italy. They assumed the electron in hydrogen was in an excited state, very close to the transition between classical and quantum mechanics. Approximating it with a classical system consisting of a particle in orbit, they considered what would happen when it absorbed a series of impulses. In the atomic case these impulses would come from an oscillating electromagnetic wave, their intensity depending on the position of the particle in its orbit, but in the classical model they had to be looked upon as arbitrary "kicks." If the system was nonchaotic no energy would be absorbed, but if it was chaotic (classically) energy would be absorbed on the average at a constant rate.

What happens when this system goes through the transition from classical to quantum mechanics? The investigators found that the two descriptions gave identical results at first—both absorbed energy in the same way. But after a time the quantum mechanical system broke away—the absorption of impulses decreased and became erratic. Quantum mechanics was obviously suppressing the classical chaos. But what about the

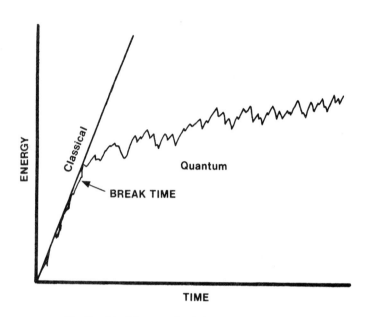

The "breaking" between classical and quantum systems.

correspondence principle that says that the two descriptions must give the same result? It turns out that it is still valid; if you make the particle heavier, which makes it more classical, the breakdown comes later.

The experimental version of this is difficult to carry out, and has never been completely done. But the experiments that have been performed indicate that there is indeed a break and a suppression of chaos in the quantum mechanical regime. Nevertheless, as we will see, there is a form of chaos in this region.

Because of its simplicity the hydrogen atom is a good place to look for chaos. Let's begin, then, by considering a hydrogen atom, with its electron in a low energy level. Placing it in a strong magnetic field, we find that the electron orbits as usual; the electromagnetic attraction between the proton and electron

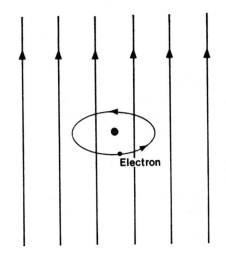

MAGNETIC FIELD

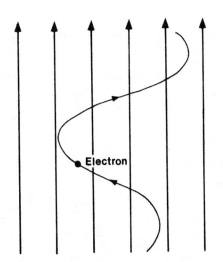

Electron in a magnetic field. Top: Field is weak and electron remains in orbit around nucleus. Bottom: Field is strong. Electron orbits field lines.

is much stronger than the force exerted by the magnetic field, and nothing strange happens. In short, there is no chaos. Furthermore, if the electron is in a highly excited orbit there is no chaos. In this case the electron is so weakly attracted to the proton that the magnetic field overcomes it and the electron spirals around the magnetic field lines. In between these two cases, however, there is a region of energies where the two forces are comparable, and it is difficult for the electron to decide which way to go. As a result it becomes chaotic.

If we look at a plot of what happens in this case we see something interesting. Chaos is present and therefore the excited energy levels of the hydrogen atom do not have any order; they are completely irregular. But let's consider the statistics of the

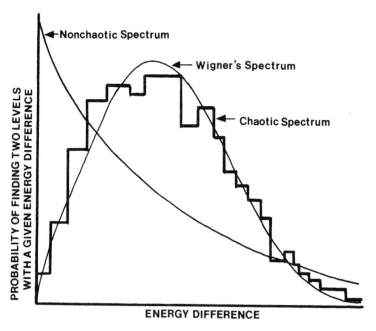

Histogram showing energy spectrum of energy levels.

energy level spacings, in other words, the statistical distribution of the spacings between neighboring levels. If we make a histogram for the case when electron orbits are chaotic we get the result shown in the figure on page 274. The solid curve that the histogram follows is one that was derived by Nobel laureate Eugene Wigner for the irregular spectra found in complex nuclei, but also applies to the energy levels of the hydrogen atom in a magnetic field. Wigner used what is called random matrix theory to derive his curve (it cannot predict the exact location of energy levels but can give a statistical estimate of the fluctuations of the spacings).

This figure shows us that the energy levels are not as random as they appear. Consider first the case where the levels are roughly equal, where they are approximately the same distance apart. The distribution in this case will cluster about some average value; this is what occurred in the hydrogen atom above. If the levels were randomly distributed there would be a large number of small separations, and histogram would follow what is called the Poisson distribution. The levels of the hydrogen atom in a magnetic field follow this curve when there is no chaos. This indicates that when the atom becomes chaotic there is a "repulsion" between the levels; they stay as far apart from one another as possible. This does not happen in the nonchaotic case.

Another place where we see a form of chaos in quantum mechanics is in the determination of the position of the electron in the hydrogen atom. The electron cannot be pinpointed as it can be in classical mechanics; it is represented by a probability cloud, and its position can only be given as a certain probability. Its orbit around the nucleus is therefore a broad smear, or cloud.

Again we can compare the classical and quantum mechanical descriptions of position for the case where the classical description exhibits chaos. Consider a ball bouncing around in a stadium-shaped box. It is well-known that if the box is regular—rectangular or circular—the pattern of trajectories is regular and predictable. For a stadium-shaped box, however, the ball bounces around chaotically as shown in the figure.

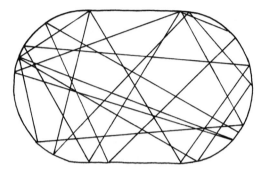

Chaotic orbits in stadium-shaped container.

Particle in stadium-shaped box. Most of the states are concentrated around narrow channels.

Poincaré section of hydrogen atom.

The stationary states (wave patterns that do not vary in time) in the quantum mechanical analog of this were calculated by Eric Heller of the University of Washington and his students in the early 1980s. His results were interesting; he found that most stationary states were concentrated in certain regions, forming strange shapes throughout the configuration. They were, in fact, similar to the stationary states of a hydrogen atom in a strong magnetic field. Although this is not a clear indication of chaos, it appears to have some properties of it. Furthermore, the Poincaré section for an electron in a hydrogen atom in a strong magnetic field shows that the electron is chaotic in certain regions.

Another place where chaos shows up in the quantum realm is in scattering, such as when the electron is scattered by several molecules (see diagram). As the electron threads its way through these molecules it can have an exceedingly complex trajectory. Small differences in its entry direction or entry energy make

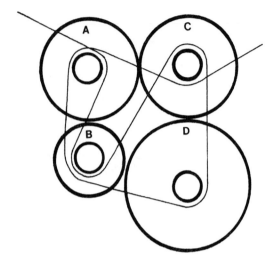

Trajectory of an electron through a molecule.

dramatic changes in its path, as well as where it exits. Furthermore, the longer the pathlength, the larger the number of possible trajectories. The path can only be calculated using quantum mechanics, and since it depends critically on initial conditions, it has characteristics of chaos.

Finally, an intriguing connection between quantum chaos and number theory has recently been uncovered. The full significance of it is not yet understood. In 1859 the German mathematician Georg Riemann was studying the distribution of prime numbers (numbers with no divisor). He defined a function called the zeta function and showed that the points where this function vanished—its zeros—lie on a line. No one has ever proven that they all lie on this line, but it is known to be true for numbers up into the billions.

Andrew Odlyzko of Bell Laboratories has shown that the distribution of spacings between neighboring zeros gives exactly the same plot as the spacings between the energy levels when there is no symmetry in a chaotic system. In fact, a theorem has recently been proved that suggests the zeta function may be able to describe all the chaotic behavior a quantum system can exhibit.

THE LARGE-SCALE STRUCTURE OF THE UNIVERSE

Now that we have some familiarity with quantum chaos we ask: How does it relate to astronomy? There are several places where it appears to be important, and one of them is the large-scale structure of the universe. In the last chapter we saw that a number of people are trying to see if this structure is related to Mixmaster action in the early universe, but as we saw, most of the Mixmaster oscillations occur within the quantum realm and are too small to be properly described by general relativity. They could be described by a quantum version of general relativity, but we have no such theory.

Most astronomers are now convinced that the large-scale structure has its origins in quantum fluctuations that occurred shortly after the creation of the universe. What role chaos played in this, if any, is uncertain, but it may have played an important role.

The first indication that the large-scale structure of the universe was inhomogeneous came in the late 1970s and early 1980s. A number of large voids—regions where there are no galaxies—were discovered in space. Furthermore, it was becoming evident that the largest known structures in the universe, superclusters (clusters of clusters of galaxies) were not uniformly distributed. James Peebles of Princeton was skeptical of the evidence; he was convinced that on a very large scale the distribution of matter was uniform. To check it out, he and a number of graduate students plotted approximately 1 million galaxies and found to their surprise that they were not uniformly distributed. A distinct mottling could be seen: some regions were densely covered, others had few or no galaxies.

But was this a valid representation? Galaxies are at many different distances. Some are close, others much farther away, and Peebles' plot was ignoring this; it was two-dimensional. Perhaps the structure was due to the way things were plotted. What was needed was a three-dimensional plot, but obtaining such a plot was a challenge. The redshift of all the galaxies would be needed, and Peebles had plotted roughly a million. Redshift is obtained from the spectrum of the galaxy (it tells us how fast a galaxy is moving away from us), and at that time it took hours to get the spectrum of a single galaxy.

Marc Davis and John Huchra, who were then at Harvard University, decided to see if they could get the required spectra. They wouldn't be able to get a million, or even hundreds of thousands, but they might be able to get enough to make a reasonable three-dimensional plot. They had access to a 60-inch reflector at Mt. Hopkins in Arizona, but it was in bad shape. Within a few months, however, they had equipped it with the latest imaging devices and could get approximately a dozen spectra a night. They decided on a initial goal of 2400 spectra, which they believed would take about two years.

By the early 1980s they had accomplished their goal. Plotting their data they saw that the mottling was still there. They planned to extend the study to dimmer galaxies, but Davis took a job elsewhere and Huchra was left on his own.

About the time Davis left, Margaret Geller returned from a post-doctoral in England, and Huchra teamed up with her. The first problem they had to face was how to extend the survey. They couldn't continue taking spectra of all galaxies, in other words, surveying all galaxies slightly dimmer than those in the earlier survey. There were too many, so they had to restrict themselves. They decided to take the spectra of all galaxies in a pie-shaped wedge in space. By the time they had completed three wedges, they saw a distinct structure beginning to emerge. Large spherical bubbles—some up to 150 million light years across—were beginning to appear. Some of them corresponded to the voids that were seen earlier, but Huchra and Geller saw a more

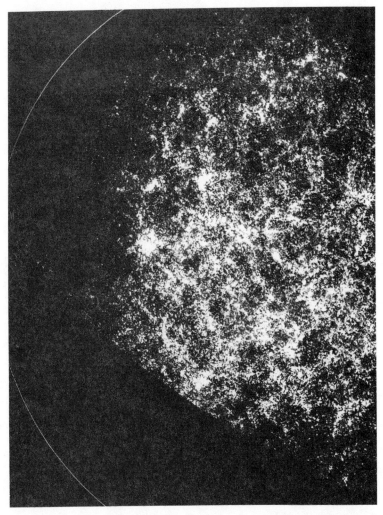

Peebles' plot of galaxies showing two-dimensional distribution. (J. Peebles)

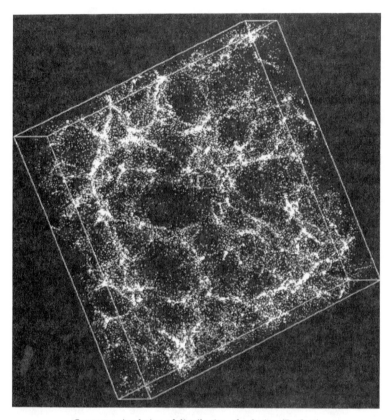

Computer simulation of distribution of galaxies. (R. Gott)

general pattern. The universe, it seemed, was made up of superclusters strung over the surface of huge bubbles. It had a Swiss cheese-like structure.

In 1989 David Koo and Richard Kron, who were then at the University of California at Berkeley, reported a much deeper probe into space, far beyond the limit of Huchra and Geller's pie slices, but it was a narrow pencil-like probe. They reported that they found "walls of galaxies," regions where the galaxies were so dense they literally constituted a wall. Other surveys reported similar structures; one was referred to as QDOT (the first initials of the universities involved: Queens, Durham, Oxford, and Toronto) and another as APM (automatic plate measuring device). The evidence now seems to be overwhelming that on a very large scale the universe has a distinct honeycomb structure made up of walls, so the problem becomes: What caused it? Several computer simulations have been made that simulate fairly well what is seen. In all cases, however, they are forced to assume the existence of a large amount of what is called dark matter. This is matter we cannot see, but believe exists because we see evidence of its gravitational effects. A large fraction of the matter of the universe, in fact, is assumed to be dark matter.

Regardless of the existence of dark matter, the major problem is still the origin of the fluctuations that created the galaxies. They had to occur at a very early stage in the formation of the universe—at a time when quantum effects were important. According to the accepted scenario, space fluctuated or "quivered" slightly. It could do this because of the uncertainty principle, which states that certain variables such as energy, time, position, and momentum can never be known exactly; they are uncertain and therefore their measured values can fluctuate slightly. The details of the process have never been worked out, and we are still uncertain whether it is correct. Could chaos have played a role in it? The universe was in a state of near chaos at the time, and since the only valid description is a quantum one it is reasonable to assume that quantum chaos might have played a role.

If we push back even further into the early universe we eventually get into a region where all known theories break down. During this time the universe was chaotic; we have no theory for describing it. John Wheeler has suggested that a "quantum foam" may have been present. It was a foam of space and nonspace—a fluctuating region filled with wormholes of spacetime. It is a small, dense region where the general theory of relativity is no longer valid. To describe it, we would need a quantum cosmology, but we have no such cosmology and so it is another place where quantum chaos may play an important role.

CHAOS AND THE THEORY OF EVERYTHING

The problems associated with the initial singularity of the universe bring us to what is called the theory of everything. It is an all-encompassing theory that would completely explain the origin of the universe and everything in it. It would bring together general relativity and quantum mechanics, and explain everything there is to know about the elementary particles of the universe, and the four basic forces of nature (gravitational, electromagnetic, weak, and strong nuclear forces). Furthermore, it would explain the basic laws of nature and the fundamental constants of nature such as the speed of light and Planck's constant.

The formulation of such a theory has been the goal of physicists for decades. They are now approaching the problem on several fronts. A theory called superstring theory that assumes the existence of tiny strings, appears to be one of the most promising approaches. All elementary particles are assumed to be made up of these strings, and the forces of nature are also explained in terms of them, but so far it has only had partial successes.

A quantum cosmology which would be an important step toward a theory of everything, if successful, has been formulated by John Wheeler of Princeton and Bryce De Witt of the University of Texas. In quantum mechanics the central equation is Schrödinger's equation; it gives the time development of the wave function representing a system. Wheeler and De Witt have come up with an equation that is analogous to Schrödinger's equation, but includes aspects of general relativity. The wave function in this case is called the "wave function of the universe." In theory if you could solve this equation you could determine anything there is to know about the universe. You could trace back to the origin of the universe—to the singularity, past the point where general relativity breaks down.

There are, unfortunately, serious difficulties with the basic equation of the Wheeler–De Witt theory. Like any differential equation you need initial conditions to solve it, but it is the initial conditions that we are interested in obtaining. So there's a catch-22. We must know what we want before we can determine it.

We have grown accustomed to the notion that everything is calculable, and that the world is deterministic. But chaos theory has shown us that this is not necessarily true. Many things appear to be beyond our mathematical models, and this means a theory of everything may not exist. Chaos may forbid us from knowing everything.

15

Epilogue

*T*his *brings us to the end of our journey into the world of chaos. We have seen that chaos theory is one of the most exciting* developments in science in the last 30 years, a development that has completely changed the way we look at nature. At one time the universe was considered to be deterministic. With enough machinery, we could follow every particle in the universe throughout its entire history. We could, in effect, calculate everything that is to be known about the universe. But this assumed that nature was linear, satisfying linear equations that were easy to solve, and scientists gradually began to realize this wasn't the case. Much, if not most, of nature is nonlinear, and with this nonlinearity comes unpredictability and chaos. Yet, strangely, the chaos is not completely random; it has structure. Chaotic trajectories don't wander randomly throughout space; they are confined.

And as we have seen, chaos theory has been applied extensively to astronomy. The gaps in the asteroid belt are now believed to be a result of chaos, and the orbits of at least one moon and several planets of the solar system are chaotic. Beyond the solar system a number of pulsating, or variable, stars have been shown to be chaotic. Furthermore, some of the stars of our galaxy and other galaxies appear to have chaotic orbits. And finally we have seen that chaos is present in general relativity, in the orbits around black holes, in cosmology, and the early universe.

So far, though, we have only scratched the surface. The application of chaos theory to astronomy is a new and blossoming field, and many other objects and areas will soon come under its influence. One is the interior of the Earth. We are still uncertain what is going on beneath the surface of our planet, but we see one of its effects in the reversal of our magnetic field every 100,000 to 150,000 years. It is an unpredictable, apparently random, event that may one day be explained by chaos theory.

Chaos also plays an important role in what is going on above the surface, namely the weather. But weather occurs on other planets. Tremendous storms rake the surface of Mars, and there is evidence for violent thunderstorms beneath the clouds of Venus and Jupiter. Chaos may one day help explain them.

It may also explain many of the features of the violent, unpredictable surface of our sun. In fact, the surfaces of all stars are likely chaotic; neutron stars, for example, are known to give rise to random bursts of x rays called bursters.

We talked in the book about chaotic orbits of stars in galaxies, but chaos may also play a role in the motions of clusters of galaxies and superclusters. Indeed, we saw that the large-scale overall structure of the universe may be a result of chaos.

Finally, there is chaos theory itself. What is in store for it? How is it likely to change over the next few years? Despite the tremendous advances that have been made, chaos theory still has defects. One of its most serious is that it does not specify the set of circumstances needed for a given sequence of events to end in chaos. In other words, it does not give explicit prerequisites for chaotic behavior. Scientists are now looking into this.

Also, what, we may ask, is beyond chaos? Considerable interest has recently been shown in what is called anti-chaos, or complexity. (I hesitate to use the word "complexity" here because it is used in so many different ways.) Complexity, or anti-chaos, can be thought of as the opposite of chaos in the following respect. Chaos shows us that simple systems sometimes produce very disorganized behavior. Complexity, on the other hand, shows us that complicated behavior, or complicated rules, sometimes give organized behavior.

Another area that is getting considerable interest lately is "the edge of chaos"—the never-never land between chaos and nonchaos. Fantastic pictures similar to those that come from the Mandelbrot and Julia sets have been obtained. But the study is more than just a study of pretty pictures; several important results have been obtained.

The phenomenon of chaos has been known for hundreds of years, but the science of chaos is still in its infancy. Scientists have only studied it seriously for the last 30 years. We are still uncertain what the future holds, but optimism is high.

Glossary

Active galaxy A galaxy that is emitting large amounts of energy from near its core. A radio galaxy.

Amplitude The maximum height above or below the zero point that a wave achieves.

Analytical mechanics Mechanics based on the use of calculus.

Angular momentum A measure of the spin of an object.

Asteroid Very small member of the solar family. Most are between Mars and Jupiter.

Astronomical unit The distance between the Earth and the Sun.

Attractor Something a dynamical system settles down to after a period of time.

Barred galaxy A spiral galaxy that has a barlike structure through its nucleus.

Bifurcation A splitting. Period doubling.

Binary galaxy A double galaxy. Two galaxies in a system.

Binary pulsar A system consisting of a black hole and a neutron star.

Black hole A region of spacetime from which nothing, not even light, can escape.

Bode's law A relationship between the orbital radii of the planets.

Burster A small star that suddenly flares up.

Cantor set Take a line, remove the center third (leaving ends). Remove the center third on each of the ends. Continue indefinitely.

Celestial mechanics The calculation of orbits of planetary and stellar systems.

Comet A small body composed of ice and dust. As it comes close to the sun it develops a tail.

Complex number Has the form $x + iy$ where x and y are real numbers and i is the imaginary unit (square root of -1).

Complex plane Plane where complex numbers are plotted.

Convergent series A series of numbers that, when summed, give a finite value.

Cosmogony Theory of the origin of the solar system, or universe.

Cosmology Theory of the structure of the universe.

Dark matter Matter astronomers cannot see but know exists in the universe.

Determinism The idea that everything in the universe is determined by specific laws.

Difference equation An algebraic equation that involves small differences in variables.

Differential equation Equations involving rates of change. Basic equation of calculus.

Dissipative A system that loses energy from friction or other means.

Divergent series A series of numbers that, when summed, give an infinite result.

Doppler effect The apparent change in wavelength of light due to relative motion between source and observer.

Eccentricity A measure of the elongation of an elliptical orbit.

Electromagnetic field One of the four basic fields of nature. The field that holds atoms together.

Electron Fundamental particle of nature.

Ellipse One of the conic curves. An egg-shaped curve.

Elliptical galaxy A galaxy that has an elliptical shape. Usually contains old stars.

Entropy A measure of the disorder of a system.

Escape velocity The velocity needed to completely escape the gravitational pull of the Earth or other object.

Euclidean geometry Geometry based on the principles of Euclid. Type of geometry taught in high school.

Event horizon Surface of a black hole. A one-way membrane.

Feynman diagram A small diagram consisting of lines, arrows, and so on, depicting collisions of elementary particles.

Field theory The theory of the interaction of particles and fields.

Fractal Something that is self-similar at all scales.

Fractal dimension The dimension of a fractal. Usually a nonintegral or fractional value.

Galactic dynamics Study of the motions of galaxies.

General relativity Theory of gravity and accelerated motion devised by Einstein in 1915.

Geodesic Shortest (or longest) distance between two points.

Globular cluster A group of a few hundred thousand stars (sometimes a few million).

Gravitational radiation Radiation caused by changes in the distribution of matter.

Ground state Lowest energy state of an electron in an atom.

Halo (of galaxy) Halo of invisible matter surrounding a galaxy.

Imaginary number An integral multiple of the square root of minus one.

Integral A mathematical operation. A summing.

Kuiper cloud A belt of comets just beyond orbit of Pluto.

Laser Device for producing coherent light (light with no internal scattering).

Limit cycle A closed loop trajectory surrounding a source.

Linear An equation is linear if the sum of two solutions is also a solution.

Local cluster The group of approximately 25 galaxies that includes the Milky Way.

Logistic equation A difference equation that is frequently used for predicting ecological populations.

Lyapunov coefficient Gives a measure of the sensitive dependence of a system on initial conditions. A positive Lyapunov coefficient signifies chaos.

Mapping A rule that gives a value in terms of known values (e.g., $x - 5x$)

Matrix An array of numbers.

Mixmaster model Model of the universe that assumes mixerlike motions, i.e., stretching and flattening.

Momentum Mass times velocity. Gives a measure of the inertia of a body.

Negatively curved space A negatively curved surface is like the surface of a saddle. Project to three dimensions.

Neutron star A star made up mostly of neutrons.

Noise The "static" in a signal. Random fluctuations.

Nonlinear Pertains to equations. When two solutions are added, they do not give a valid solution to the equation.

Nuclear reaction An energetic reaction involving changes in nuclei.

Number theory A branch of mathematics that deals with properties of numbers.

Oort cloud A cloud of comets about one light year from the sun.

Orrery Mechanical model of the solar system.

Parabola One of the conic curves. An open curve similar to the reflecting surface in the headlight of a car.

Perihelion The distance of closest approach between the sun and a planet in an elliptical orbit.

Perimeter The distance around a circle or other area.

Perturbation A disturbance that causes a small change in the system.

Phase point A point in phase space. Represents the system at an instant of time.

Phase space Space used in describing the state of a system.

Photon A particle of light.

Planck's constant A fundamental constant of nature designated by h. Important in quantum theory.

Polynomial equation An equation involving powers of an unknown, usually designated as x.

Positively curved space The surface of a ball has positive curvature. An extension of it to three dimensions.

Proton Fundamental particle of nature. Has positive charge.

Quadratic equation An equation involving the square of x.

Quantum mechanics Theory of the microcosm.

Quasiperiodic Almost periodic, but not quite.

Radiation Electromagnetic energy. Photons.

Redshift A shift of spectral lines toward the red end of the spectrum. Indicates recession.

Renormalization A method of readjusting values within a theory.

Simultaneous equations Sets of equations in more than one variable.

Singularity (of universe) A point of no dimensions. Initial state of the universe.

Spectral lines Lines obtained when light is passed through a spectroscope.

Spectroscope Instrument that separates the different frequencies of light.

Spin–orbit resonance When the spin period and orbital period are integral multiples of one another.

Spiral galaxy A galaxy with spiral arms.

State of system A specification of the position and velocity of all particles in a system.

Stellar dynamics Study of the motion of stars.

Strange attractor Attractor associated with chaos. Signifies chaos.

Supercluster A cluster of clusters of galaxies.

Superstring theory Theory that assumes that all particles of nature are composed of tiny strings.

Tide-locked Held together as a result of a tidal force. Our moon is tide-locked to the Earth.

Topology Branch of mathematics. A study of the properties of surfaces.

Torus Surface like the surface of a tire.

Trajectory Orbit.

Uncertainty principle Basic principle of quantum mechanics. Says that on the atomic level there is a "fuzziness" associated with nature.

Variable star Star that changes in brightness.

Velocity Speed in a particular direction.

Vortex Turbulent region. Whirling fluid.

Wave function Used in quantum theory to designate state of a system.

Bibliography

The following is a list of general and technical references for the reader who wishes to know more about the subject. References with an asterisk are of a more technical nature.

CHAPTER 1

Gleick, James, *Chaos: Making a New Science* (New York: Viking, 1987).
Stewart, Ian, *Does God Play Dice?* (Cambridge, Mass.: Blackwell, 1989).

CHAPTER 2

Bell, E. T., *Men of Mathematics* (New York: Simon and Schuster, 1937).
Ferris, Timothy, *Coming of Age in the Milky Way* (New York: Doubleday, 1988).
Littmann, Mark, *Planets Beyond: Discovering the Outer Solar System* (New York: Wiley, 1990).
More, Louis, *Isaac Newton* (New York: Dover, 1962).
Peterson, Ivars, *Newton's Clock: Chaos in the Solar System* (New York: Freeman, 1993).

Stewart, Ian, *Does God Play Dice?* (Cambridge, Mass.: Blackwell, 1989).
Westfall, Richard, *Never at Rest: A Biography of Isaac Newton* (Cambridge: Cambridge University Press, 1980).

CHAPTER 3

Bronowski, J., *The Ascent of Man* (Boston: Little, Brown, 1973).
*Holmes, P., "Poincaré, Celestial Mechanics, Dynamical Systems and Chaos" *Physics Reports* 193 (September 1990) 137.
Peterson, Ivars, *Newton's Clock: Chaos in the Solar System* (New York: Freeman, 1993).
Poincaré, Henri, *Science and Hypothesis* (New York: Dover, 1952).
Stigler, Stephen, *The History of Statistics* (Cambridge: Belknap, 1986).

CHAPTER 4

*Berge, Pierre, Pomeau, Yves, and Vidal, Christian, *Order Within Chaos* (New York: Wiley, 1984).
*Cambel, A., *Applied Chaos Theory* (New York: Academic Press, 1993).
Peak, David, and Frame, Michael, *Chaos Under Control* (New York: Freeman, 1994).
Stewart, Ian, *Does God Play Dice?* (Cambridge, Mass.: Blackwell, 1989).
Tritton, David, "Chaos in the Swing of a Pendulum" *New Scientist* (July 24, 1989) 46.

CHAPTER 5

*Devaney, Robert, *A First Course in Chaotic Dynamic Systems* (Reading: Addison Wesley, 1992).
Dresden, Max, "Chaos: A New Scientific Paradigm" *The Physics Teacher* 30 (January 1992) 10.
Kellert, Stephen, *In the Wake of Chaos* (Chicago: University of Chicago Press, 1993).
*Ott, Edward, *Chaos in Dynamical Systems* (Cambridge: Cambridge University Press, 1993).

Peterson, Ivars, *The Mathematical Tourist* (New York: Freeman, 1988).
Ruelle, David, *Chance and Chaos* (Princeton: Princeton University Press, 1991).

CHAPTER 6

Crutchfield, James, Farmer, J. Doyne, Packard, Norman, and Shaw, Robert, "Chaos," *Scientific American* 226 (December 1986) 46.
Gleick, James, *Chaos: Making a New Science* (New York: Viking, 1987).
Peak, David and Frame, Michael, *Chaos Under Control* (New York: Freeman, 1994).
Ruelle, David, *Chance and Chaos* (Princeton: Princeton University Press, 1991).
Shaw, R., *The Dripping Faucet as a Model Chaotic System* (Santa Cruz: Ariel Press, 1984).

CHAPTER 7

Barnsley, Michael, *Fractals Everywhere* (New York: Academic Press, 1988).
*Falconer, Kenneth, *The Geometry of Fractal Sets* (Cambridge: Cambridge University Press, 1985).
*Feder, Jens, *Fractals* (New York: Plenum, 1988).
Mandelbrot, Benoit, *The Fractal Geometry of Nature* (New York: Freeman, 1983).
Peterson, Ivars, *The Mathematical Tourist* (New York: Freeman, 1983).
Peitgen, H., and Richter, P., *The Beauty of Fractals* (Berlin: Springer-Verlag, 1986).

CHAPTER 8

Hartley, Karen, "Solar System Chaos" *Astronomy* 18 (May 1990) 34.
Kerr, Richard, "Does Chaos Permeate the Solar System?" *Science* 224 (April 14, 1989) 144.

Killian, Anita, "Playing Dice with the Solar System" *Sky and Telescope* 78 (August 1989) 136.

Murray, Carl, "Is the Solar System Stable?" *New Scientist* 124 (November 24, 1989) 60.

Peterson, Ivars, *Newton's Clock: Chaos in the Solar System* (New York: Freeman, 1993).

CHAPTER 9

Binzel, Richard, Barucci, Antonietta, and Fulchignoni, Marcello, "The Origin of the Asteroids" *Scientific American* 265 (October 1991) 88.

Cunningham, Clifford, "The Captive Asteroids" *Astronomy* 20 (June 1992) 40.

Gehrels, Tom, and Matthews, Mildred, *Asteroids* (Tucson: University of Arizona Press, 1979).

Hartley, Karen, "Solar System Chaos" *Astronomy* 18 (May 1990) 34.

Killian, Anita, "Playing Dice with the Solar System" *Sky and Telescope* 78 (August 1989) 136.

Murray, Carl, "Earthward Bound from Chaotic Regions of the Asteroid Belt" *Nature* 315 (June 27, 1985) 712.

Saha, Prasenjit, "Simulating the 3:1 Kirkwood Gap" *Icarus* 100 (December 1992) 434.

Wisdom, Jack, "Meteorites May Follow a Chaotic Route to Earth" *Nature* 315 (June 27, 1985) 731.

Wisdom, Jack, "Urey Prize Lecture: Chaotic Dynamics in the Solar System" *Icarus* 72 (November 1987) 241.

CHAPTER 10

Binzel, Richard, Green, Jacklyn, and Opal, Chet, "Chaotic Rotation of Hyperion?" *Nature* 320 (April 10, 1985) 511.

Klavetter, James, "Rotation of Hyperion" *The Astronomical Journal* 97 (February 1989) 570.

Murray, Carl, "Chaotic Spinning of Hyperion" *Nature* 311 (October 25, 1984) 705.

Wisdom, Jack, Peale, Stanton, and Mignard, Francois, "The Chaotic Rotation of Hyperion" *Icarus* 58 (May 1984) 137.

CHAPTER 11

Kerr, Richard, "From Mercury to Pluto, Chaos Pervades the Solar System" *Science* 257 (July 3, 1992) 33.

Laskar, J., "A Numerical Experiment on the Chaotic Behavior of the Solar System" *Nature* 338 (March 16, 1989) 237.

Laskar, J., and Robutel, P., "The Chaotic Obliquity of the Planets" *Nature* 361 (February 18, 1993) 608.

Lecar, Myron, Franklin, Fred, and Murison, Marc, "On Predicting Long-term Orbital Instability: A Relation Between the Lyapunov Time and Sudden Orbital Transitions" *The Astronomical Journal* 104 (September 1992) 1230.

Quinn, Thomas, Tremaine, Scott, and Duncan, Martin, "A Three Million Year Integration of the Earth's Orbit" *The Astronomical Journal* 101 (June 1991) 2287.

Sussman, Gerald, and Wisdom, Jack, "Chaotic Evolution of the Solar System" *Science* 257 (July 3, 1992) 56.

Wisdom, Jack, "Urey Prize Lecture: Chaotic Dynamics of the Solar System" *Icarus* 72 (November 1987) 241.

CHAPTER 12

Buchler, Robert, and Regev, Oded, "Chaos in Stellar Variability" The Ubiquity of Chaos (Washington, D.C.: American Association for the Advancement of Science, 1990) 218.

Cannizzo, John, and Goodings, D., "Chaos in SS Cygni?" *The Astrophysical Journal* 334 (November 1988) L31.

*Caranicolas, N. D., and Innanen, K. A., "Chaos in a Galaxy Model with Nucleus and Bulge Components" *The Astronomical Journal* 102 (October 1991) 1343.

*Carlberg, R. G., and Innanen, K. A., "Galactic Chaos and the Circular Velocity at the Sun" *The Astrophysical Journal* 94 (September 1987) 666.

*Hasan, Hashima, and Colin, Normman, "Chaotic Orbits in Barred Galaxies with Central Mass Concentrations" *The Astrophysical Journal* 361 (September 1990) 69.

*Lochner, James, Swank, J., and Szymkowiak, A., "A Search for a Dynamical Attractor in Cygnus X-1" *The Astrophysical Journal* 337 (February 1987) 823.

Norris, Jay, and Matilsky, Terry, "Is Hercules X-1 a Strange Attractor?" *The Astrophysical Journal* 346 (November 15, 1989) 912.

CHAPTER 13

*Barrow, John, "Chaos in Einstein's Equations" *Phys. Rev. Lett.* 46, 15 (April 1981) 963.

*Berger, Beverly, "Numerical Study of Initially Expanding Mixmaster Universes" *Class Quantum Gravity* 7 (1990) 203.

*Bombelli, Luca, and Calzetta, Esteban, "Chaos around a Black Hole" *Class Quantum Gravity* 9 (1992) 2573.

*Chandrasekhar, S., "The Two-Centre Problem in General Relativity: The Scattering of Radiation of Two Extreme R. N. Black Holes" *Proc. R. Soc. London A* 421 (1989) 227.

*Contoupolos, G., "Periodic Orbits and Chaos around Two Black Holes" *Proc. R. Soc. London A* 431 (1990) 183.

*Misner, Charles, "Mixmaster Universe" *Phys. Rev. Lett.* 22 (May 1969) 1071.

*Rugh, Svend, "Chaos in the Einstein Equation: Characterization and Importance" *Deterministic Chaos in General Relativity*, D. Hobill (ed.) (New York: Plenum, 1994).

CHAPTER 14

Gutzwiller, M., "Quantum Chaos" *Scientific American* 226 (January 1992) 78.

Jensen, Roderick, "Quantum Chaos" *The Ubiquity of Chaos* (Washington, D.C.: American Association for the Advancement of Science, 1990) 98.

Index